AMGUEDDFA GENEDLAETHOL CYMRU

NATIONAL MUSEUM OF WALES

THE FLORA OF
MONMOUTHSHIRE

BY

A. E. WADE

CARDIFF

NATIONAL MUSEUM OF WALES

PUBLISHED 1970

Printed by Qualitex Printing Limited, Cardiff

PREFACE

This work is a departure from the series of Museum publications dealing with the botany of Wales, in that it is the first publication to deal with the plants of a single county rather than the flora of the country as a whole.

The author, Mr. A. E. Wade, was Assistant Keeper in the Department of Botany until he retired in 1961 after 42 years' service in the National Museum. His botanical explorations in Monmouthshire have extended over 45 years, yet he makes no claim that the work is exhaustive, even though it is the first to bring together from a variety of sources a full account of the flowering plants, ferns and fern allies in the county. The need for an adequate Flora is apparent when previous works on the plants of the county are reviewed.

A Flora of Monmouthshire by J. H. Clark was published in 1868 and another by Samuel Hamilton in 1909. Both of these Floras were based almost entirely on personal records and included less than 700 species. W. A. Shoolbred's *Flora of Chepstow* (1920) covered more ground than the title implies, listing 1,004 vascular plants. This was a remarkably large number for an area extending only 9 miles along the lower reaches of the Wye and 5 to 7 miles from each bank. A few species from outside this area were included, for their rarity or interest, but there was no suggestion that the *Flora of Chepstow* was in fact a Flora of Monmouthshire.

The present Flora covers the whole of the county. One thousand, one hundred and seventy-four species are listed of which 37 are Pteridophyta, 2 Gymnospermae and 1,135 Angiospermae. Of the Pteridophyta, 36 are native and one naturalized. Of the Gymnospermae one is native and the other naturalized. Of the Angiospermae 823 are native, 7 doubtfully native, 94 denizens, 55 colonists, 90 aliens and 66 casuals (for definitions of these categories, see p.42).

The author has been greatly indebted to Mr. D. Emlyn Evans, who provided the account of the geology of the county.

The Keeper of Botany, Mr. S. G. Harrison, has been responsible for arranging the publication of this work and for reading the text on behalf of the National Museum. The Assistant Keeper, Mr. R. G. Ellis, has helped in many ways. Thanks are also due to the Monmouthshire Naturalists' Trust for their interest in the project and to those members who provided transport and accompanied the author on field excursions, to Mrs. J. D. Davies for assistance in typing the manuscript, Mr. J. D. Davies, Mr. R. L. Smith who assisted in the early stages of the work, and the Controller of H.M. Stationery Office for permission to reproduce the map of Monmouthshire.

Grateful thanks are due to the many correspondents who have contributed records, and especially to Mr. S. G. Charles, Mr. T. G. Evans, Miss B. M. Frederick and Mr. A. McKenzie for lists of plants observed in their neighbourhoods, and to Mr. E. S. Edees for the list of Rubi. The following have determined specimens of critical genera: Dr. C. D. K. Cook (*Ranunculus* subgen. *Batrachium*), Dr. D. H. Dalby (*Salicornia*), Mr. J. E. Dandy (*Potamogeton*), Dr. R. Melville (*Ulmus*), Dr. T. D. Pennington (*Epilobium*), Mr. V. S. Summerhayes (*Dactylorhiza*), Sir George Taylor (*Potamogeton*), the late Dr. E. F. Warburg (*Sorbus*), and Dr. P. F. Yeo (*Euphrasia*).

The purpose of a County Flora is to provide a guide to the distribution of plants within the county in much greater detail than can be expected in a National Flora. It is hoped and expected that it will provide a stimulus to both local and visiting naturalists to add to our existing knowledge of the changing plant population in a changing environment. A full knowledge of the plants of an area is essential for defining the need for conservation in particular localities.

G. O. JONES,
Director.

NATIONAL MUSEUM OF WALES,
CARDIFF. *July* 1970.

CONTENTS

		Page
INTRODUCTION	.. ..	9
History of the Study of the Monmouthshire Flora	..	9
The Botanical Districts	.. ..	17
The Geology of Monmouthshire by D. Emlyn Evans	..	22
BIBLIOGRAPHY	.. ..	35
ANNOTATED FLORA AND MANUSCRIPTS	.. .. ..	38
LIST OF RECORDERS ..	.. ..	39
HERBARIA	.. ..	42
PLAN OF FLORA	.. ..	42
ACCOUNT OF THE SPECIES	.. ..	45
Pteridophyta	.. ..	45
Spermatophyta	.. ..	51
INDEX	.. ..	227

LIST OF ILLUSTRATIONS

Plates

PLATE I River Wye above Chepstow	.. ..	facing page 20
PLATE II The Rhymney estuarine clay flats	.. .. ,,	21

Maps

Geological Map	.. ..	.. *at end of book*
Topographical Map showing the Botanical Districts		*at end of book*

INTRODUCTION

HISTORY OF THE STUDY OF THE MONMOUTHSHIRE FLORA

The earliest records for the county are in John Parkinson's *Theatrum botanicum*, 1640, where *Saxifraga aurea* (*Chrysosplenium oppositifolium*) is recorded from Chepstow, *Malva arborea marina nostras* (*Lavatera arborea*) from Denny Island and *Alchemilla major vulgaris* (*Alchemilla vulgaris* aggr.) is said to grow "in pastures nigh Tidnam [Tidenham] and Chepstow".

In 1662 John Ray visited Wales. Ray's diary of the visit was edited by Edwin Lankester and included in his *Memorials of John Ray*, published in 1846. On his itinerary Ray passed through Monmouthshire and he records under Saturday June 14th, "We rode to Caerwent . . . near Creek [Crick] in a bushet or wood on an (*sic*) hill, not far from the wayside, westward, grows *Vicia maxima sylvatica* (*Vicia sylvatica*) *forte et erythrodamum* (*Rubia peregrina* L.) besides many other common plants . . . by the way, near Tintern, I found Herba Paris (*Paris quadrifolia*) and a kind of *Driopteris* q. an *Driopt. Tragi?*" (*Gymnocarpium dryopteris*). The record for *Gymnocarpium dryopteris* is given by Ray in his *Catalogus Plantarum Angliae*, 1670, and in his *Synopsis Methodica Stirpium Britannicum*, ed 1, 1690; that for *Vicia sylvatica* is also referred to in the latter.

Edward Lhwyd (1660-1709) discovered *Anaphalis margaritacea* which is reported in Ray's *Synopsis*, ed. 3, 1724 as growing "on the banks of the Rymny River, for the space of at least twelve miles".

The Rev. John Lightfoot (1735-1788) visited Wales in the company of Sir Joseph Banks in 1773. A transcript made by Solander of Lightfoot's journal of the visit is in the Department of Botany, Natural History Museum, South Kensington. This was edited by the late H. J. Riddelsdell and published in the *Journal of Botany*, vol. 43, 1905.

Under Sunday, June 27th, Lightfoot records *Alopecurus bulbosus* in "the marshes by the Severn side going the Foot way from the new Passage House to Chepstow" and *Brassica oleracea* upon Chepstow Castle. In Piercefield Woods were found *Hordelymus europaeus*, *Agropyron caninum*, *Bromus madritensis*, *Rubia peregrina*, *Melica nutans*, var.; *Sedum forsterianum* ssp. *elegans* on rocks by the Wye, at Chepstow; *Ophrys apifera* and *O. insectifera* on steep banks by the Wye, near Chepstow Castle; *Cephalanthera longifolia* and *Geranium sanguineum* in the Castle Wood, Chepstow and on the Wynd Cliff.

Under Tuesday, June 29th, is recorded *Euphorbia serrulata* (as *E. platyphyllos*) "by the Brook Side, going from the Abbey (Tintern) to the Forge where they make wires", and *Mentha rotundifolia* near the same place.

Entering Christchurch on Thursday, July 1st, they saw *Leonurus cardiaca*. On Monday, July 5th, they set out from Cardiff to visit the Rumney marshes where they "saw large crops of the *Plantago maritima*

9

INTRODUCTION

call'd here by the people *Gibbals*, which the Hogs are very fond of . . . In the same marshes we saw abundance of the *Alopecurus bulbosus* and *Hieracium paludosum*". [This latter record, if *Crepis paludosa* is intended, is certainly an error]

Richard Warner (1763-1857), divine and antiquary, born in Marylebone, London, was a voluminous writer on topography, antiquities and theology. In 1797 he published a very popular work entitled *A Walk through Wales in August 1797:* subsequent editions appeared in 1798 and 1799. In it he remarks upon the profusion of *Artemisia absinthium* growing near Caldicot Castle.

William Pitt (1749-1823), a writer on agriculture, contributed some records to the 4th edition of Withering's *Arrangement of British Plants*, 1801, including a record of *Galium odoratum* from the Chepstow district.

George William Manby (1765-1854), in his *Historic and picturesque guide . . . through the counties of Monmouth, Glamorgan and Brecknock*, published in 1812, records *Sambucus ebulus* from near the walls of Caerwent, where it is still to be found.

Edward Donovan (1768-1837), in *Descriptive excursions through South Wales and Monmouthshire in the year 1804 and the four preceding years*, published in 1805, repeats Manby's record for *Sambucus ebulus* and gives the additional localities of Caerleon and Raglan Castle.

The Botanist's Guide by Dawson Turner and L. W. Dillwyn, published in 1805, contains a brief list of but 10 plants including three mosses. Only one flowering plant is new to the county, *Hypericum androsaemum* on the authority of Joseph Woods. Sir Joseph Banks is given as the authority for the record of *Brassica oleracea* from below Chepstow Castle, which he saw in Lightfoot's company. Sir Thomas Gery Cullan is quoted as responsible for the record of *Rubia peregrina* from Piercefield, but he probably had the record from Banks or Lightfoot since there is no evidence that he ever visited Monmouthshire.

A copy of *The Botanical Pocket Book* by William Mavor published in 1800, now in the possession of Mr. J. S. L. Gilmour, contains a number of MS records. Twenty-one of these were made in Monmouthshire in the years 1806 and 1807, and all but one are new records for the county. Unfortunately the recorder is not known.

Thomas Purton (1768-1833) includes Monmouthshire in his *Midland Flora*, 1817, and gives for the county *Chaenorhinum minus, Mentha rotundifolia, Origanum vulgare* and *Cystopteris fragilis*. His record for *Circaea alpina*, said to grow in woods in the neighbourhood of Abergavenny, doubtless refers to *Circaea intermedia* which occurs abundantly in a wood near Llanelen.

Thomas Walford (1752-1833) published in 1818 *The Scientific Tourist through England, Wales and Scotland*. In this work are noted "the principal

INTRODUCTION

objects of Antiquity, art, science, and the picturesque" under the heads of the several counties. The list of plants under Monmouthshire contains 12 species, mainly derived from previous writers. *Circaea lutetiana*, *Orobanche major* and *Chenopodium vulvaria*, all from Chepstow Castle, are the only additions to the county list. The record for *Orobanche major* is probably an error for *O. hederae*. The records were presumably made by Walford himself, when he visited the county in 1781.

The New Botanist's Guide, by Hewett Cottrell Watson (1804-1881), repeats in vol. 1, 1835, the short list of Monmouth plants given in Turner and Dillwyn's *Botanist's Guide* and adds records accredited to Nathaniel John Winch (1768-1838), Edwin Lees (1800-1887), Richard Chandler Alexander Prior (1809-1902), the Rev. J. Pool and the Rev. John Collins (fl. 1816-1848). The additions to the county list are *Chrysosplenium alternifolium* and *Campanula patula* recorded by E. Lees, and *Umbilicus rupestris* and *Centranthus ruber* by N. J. Winch. Volume 2, 1837, of Watson's work contains, as a supplement, a list of plants observed by Charles Conway in the neighbourhood of Pontnewydd Works. This is the first substantial contribution to the botany of Monmouthshire. Seventy-five of the 83 plants listed are new county records. The "neighbourhood" of Pontnewydd Works should be interpreted widely as the list includes maritime species and some records which were almost certainly made in the Wye Valley, e.g. *Convallaria majalis*. The list may have been supplied to Watson as a general one of Monmouthshire plants and given the title printed, under the impression that the list was confined to plants growing about Conway's place of business. This list was used as the basis of the majority of the records for Monmouthshire in the first edition of H. C. Watson's *Topographical Botany*, 1873-1874.

Charles Conway (*c*. 1797-1860) of Pontrhydyrn formed a herbarium, mainly of his own collecting during the 1830's, now in the Welsh National Herbarium at the National Museum of Wales. It contains a large number of Monmouthshire plants.

In 1842 Edwin Lees (1800-1887) published his popular work *The Botanical Looker-Out*. References to Monmouthshire plants deal chiefly with those of the Wye Valley and the Abergavenny district. The following are additions to the county list: *Sorbus aria, Crataegus monogyna* (referring to the several fine hawthorns growing on the summit of the Ysgyryd Fach, Abergavenny), *Erica cinerea, Orobanche hederae* (as *O. minor*, corrected in subsequent editions), *Antirrhinum majus* and *Athyrium filix-foemina*. The second edition, which appeared in 1851, added *Helleborus viridis, Lobularia maritima, Arabis hirsuta, Saponaria officinalis, Acer campestre, Epilobium angustifolium, Ilex aquifolium, Smyrnium olusatrum, Inula helenium, Artemisia maritima, Hyoscyamus niger, Mentha* × *verticillata, Melica nutans, Asplenium adiantum-nigrum* and *Taxus baccata*.

11

INTRODUCTION

The Rev. William Henry Purchas (1823-1903) co-author with the Rev. Augustin Ley of the *Flora of Herefordshire*, frequently botanised in Monmouthshire. Amongst his earliest finds are *Carex digitata* and *C. montana* from the Wynd Cliff reported in the *Botanical Gazette*, vol. 1, 1849.

To the same volume of the *Botanical Gazette*, John Ball (1818-1889) contributed an article entitled "Contributions to the Flora of South Wales." The Monmouthshire records (12 in number) were made in the neighbourhood of Llanthony Abbey. The following are additions to the county flora: *Hypericum maculatum, Saxifraga hypnoides, Silaum silaus, Vaccinium vitis-idaea* and *Mentha viridis*.

The British Flora by Sir William Jackson Hooker, ed. 5, 1842, records *Trifolium squamosum* from Newport on the authority of J. E. Bowman, an addition to the county flora. There is no evidence that Bowman visited Monmouthshire and it is probable that he had the record from Charles Conway, with whom he appears to have exchanged plants. The sixth edition of *The British Flora*, 1850, published under the joint authorship of Sir W. J. Hooker and George A. Walker-Arnott, makes two additions to the recorded plants of Monmouthshire, *Cardamine impatiens* and *Epilobium lanceolatum*, the latter on the authority of the Bristol botanist George Henry Kendrick Thwaites (1811-1882). No authority for the record of the former is given, but it was probably also due to Thwaites.

Joseph Woods (1776-1864), already mentioned in connection with the *Botanist's Guide*, 1805, contributed to the *Phytologist*, vol. 3, 1850, "Botanical Notes, the result of a visit to Glamorgan and Monmouthshire, in . . . 1850", which gives 10 species for the county, including the following: *Rubus caesius, R. corylifolius, R. rhamnifolius, R. rudis, R. villicaulis,* the first records of brambles for the county, *Valeriana officinalis* and *Valerianella rimosa*.

The Rev. Fenton John Anthony Hort (1828-1892) was a correspondent of Hewett Cottrell Watson to whom he sent records of a few Monmouthshire plants. He contributed to the *Botanical Gazette*, vol. 3, 1851, a short paper on *Euphorbia serrulata* (as *E. stricta*) and *E. platyphyllos*.

In the *Phytologist*, vol. 4, 1852, F. J. A. Hort wrote "Notes on the occurrence of *Aconitum napellus* and *Orobanche coerulea* in Monmouthshire".

Thomas Walter Gissing (1829-1870) visited Monmouthshire and contributed "Notes of a Botanical Excursion down the Wye" to the *Phytologist*, vol. 4, 1853. Some 35 plants from the county are mentioned, of which 24 are additions. Most of the records were made in the neighbourhood of Tintern and Raglan. Several of these are of particular interest and include *Eryngium campestre* (adventive), *Neottia nidus-avis, Coeloglossum viridis, Polygonatum odoratum* and *Convallaria majalis*.

INTRODUCTION

J. H. Clark was born in Gloucester in 1818. He set up in business as a printer at Usk in 1834 and subsequently published several "guides" of which he was the author. In 1853 he published a guide to *Cardiff and its Neighbourhood*, which contains a list of plants, including a few records from South-west Monmouthshire. In 1856 he brought out *Usk and its Neighbourhood*, with a list appended of some of the scarcer plants found within 12 miles of Usk. The list gives 19 species, nine being additions to the county list.

Clark's *Sketches of Monmouthshire* appeared in 1868 with an appendix which represents the first effort to write a complete flora of the county. This was reprinted as a pamphlet, undated, under the title of *The Flora of Monmouthshire*. It enumerates 671 flowering plants and 22 ferns and fern allies. Clark's herbarium is in the Newport Museum. Although a number of his records are substantiated by specimens in this collection, 27 are shown to be errors. These include, however, many plants known to occur in the county but which, so far as Clark's *Flora* is concerned, were based upon mis-identifications.

James Bladon (fl. 1842-1857) of Pontypool, in the *Phytologist* vol. 2, 1857, gave a new record for *Aconitum anglicum* Stapf. (as *A. napellus*) for Llangattock Lingoed. Bladon was a very critical botanist and it is unfortunate that so far as the flora of Monmouthshire is concerned the whereabouts of part of his herbarium is unknown. Although some 372 of his specimens are in the Bailey Herbarium at Manchester Museum, very few are from Monmouthshire.

Thomas Henry Thomas (1839-1915) was born at Pontypool. He took up art as a career, studying at the Royal Academy school and afterwards in Paris and Rome. On his return from Italy in 1864 he settled down in London as a painter and special artist to *The Graphic* and *The Daily Graphic*. Eventually he came to Cardiff and lived with his parents. His interests were wide and he became well known in South Wales not only as an artist but also as an amateur archaeologist and naturalist. He formed a collection of plants which contains many collected in Monmouthshire. This collection is now in the Welsh National Herbarium, at the National Museum of Wales. Thomas was especially interested in ferns and corresponded with the pteridologist, Thomas Moore, who quotes a number of Thomas's records in his works, including *Botrychium lunaria*, *Osmunda regalis* and *Thelypteris phegopteris*. Records made by T. H. Thomas also appear in *Our Native Ferns* by E. J. Lowe, published in 1867 and 1869. He contributed to the *Transactions of the Cardiff Naturalists' Society*, vol. 12, 1880, "Notes on some fine specimens of oak, yew, elm and beech, chiefly in the counties of Glamorgan and Monmouth".

Charles Cardale Babington (1808-1895), the fifth Professor of Botany at Cambridge University from 1861 to 1895, paid frequent visits

INTRODUCTION

to Monmouthshire between the years 1834 and 1876. The study of the Rubi brought Babington into constant touch with the Rev. F. J. A. Hort with whose family he stayed at Hardwick Hall, near Chepstow, in 1834. Babington on several occasions stayed with Professor Edward A. Freeman, the historian, at Llanrumney Hall. Some of the Rubi which he collected on these occasions are referred to in his monograph on *The British Rubi* published in 1869. An entry in his journal under July 29th, 1858 relates that they (he and Prof. Freeman) "went by the side of the river Rumney up as far as the bridge below Cefn Mably; saw in abundance *Antennaria margaritacea* (*Anaphalis margaritacea*) on both sides of the river at intervals, also *Saponaria officinalis*". The latter is still to be met with, but the first seems to have died out. His nephew the Rev. Churchill Babington (1821-1887) also visited Monmouthshire and sent records or specimens to H. C. Watson who included them in his *Topographical Botany*.

From 1868 the members of the Woolhope Field Naturalists' Club made excursions into the county and the records of the more interesting plants met with are frequently mentioned in the Club's reports. Amongst those who took part in the excursions were Edwin Lees, Augustin Ley, Dr. Bull and Prof. C. C. Babington. Dr. Bull contributed to the *Transactions* for 1870 an account of the mistletoe-oak at Llangattock Lingoed.

The Rev. Augustin Ley (1842-1919) contributed a large number of Monmouthshire records to the *Reports of the Botanical Record Club* for the years 1873-1886, many of these being of special interest, e.g. *Rhynchospora alba* at Trelleck Bog, *Pyrola secunda* and *Festuca sylvatica* from the Wynd Cliff, and *Polygonum mite* from the banks of the River Wye. From 1872 onwards Ley contributed many notes to the *Journal of Botany*. His herbarium, now in the Department of Botany, Birmingham University, is especially rich in Monmouthshire plants.

The first edition of *Topographical Botany* by H. C. Watson, printed privately in two volumes in 1873-1874, gives 33 species previously unrecorded for the county. Authorities for these records were E. Lees, B. M. Watkins, F. J. A. Hort, Charles Conway and the Rev. W. H. Purchas. The second edition published in 1883 added a further 21, mostly attributed to F. J. A. Hort, B. M. Watkins, C. C. Babington and the Rev. A. Ley.

An important contribution to the botany of the county was made by Augustin Ley in the *Transactions of the Woolhope Field Naturalists' Club* for 1883-1885 (1890) under the title "The Botany of the Honddu and Grwyne Valleys". The characteristic and rarer plants are listed and the flora of the two valleys contrasted.

The Rev. H. P. Reader (1850-1929) visited the Chepstow and Usk districts and contributed to the *Journal of Botany*, vol. 24, 1886, a short note on the records made in Gloucestershire and Monmouthshire on that occasion. Ten plants are given for the latter county, of which only

INTRODUCTION

Bidens cernua was new, most of the others having been already recorded by Clark.

In 1886 was published John White's *Guide to the Town and Neighbourhood of Abergavenny*, to which chapters on the geology and natural history had been "furnished by a gentleman, long resident in the locality". It seems very probable that these chapters were supplied by J. H. Clark of Usk. Ninety-seven species of flowering plants and 9 ferns and fern allies are listed, no localities are given, but all are said to be found within walking distance of Abergavenny. There are a few obvious errors and some doubtful records, although if the range of the walk is taken to be a pretty long one the latter are probably correct.

Harold Warren Monington (1867-1924) contributed in 1888 to the *Journal of Botany* records for *Vinca minor* and *Erysimum cheiranthoides* from near Tintern. He also contributed a few records to Shoolbred's *Flora of Chepstow*.

In 1889 the Woolhope Field Naturalists' Club published *The Flora of Herefordshire* under the joint editorship of the Rev. W. H. Purchas and the Rev. Augustin Ley. The Ffwddog on the west of the Honddu Valley formed at that time a detached portion of Herefordshire. This is now included in the county of Monmouthshire. In addition to this the editors included the intervening and Monmouthshire part of the Honddu Valley in their botanical district 14. In consequence a considerable number of records given in the 'Flora' for this district belong to Monmouthshire. Eighteen species are additional to those previously recorded for Monmouthshire and include *Geranium sylvaticum*, *Salix purpurea*, *Gymnocarpium robertianum* and *Lycopodium selago*. A supplement to this Flora was published in the *Transactions* of the Woolhope Club in 1894. Again this includes a number of records from our county.

William Whitwell (1839-1920) contributed Monmouthshire records to the *Report of the Botanical Record Club for* 1881-1882 and to the *Journal of Botany*, vol. 29, 1891, a short list of 9 flowering plants and one stonewort.

W. A. Shoolbred (1852-1928), a medical practitioner of Chepstow, contributed to the *Journal of Botany*, vol. 32, 1894, a long and valuable article entitled *Recent Additions to the Flora of West Gloucester and Monmouth* which considerably increased the recorded flora. In the same volume of the Journal a further contribution was added, to be followed in 1896, by *Plants of West Gloucester and Monmouth* and in 1898 by *West Gloucester and Monmouth Plants*. W. A. Shoolbred and the Rev. E. S. Marshall spent much time in each other's company on botanical expeditions and they collaborated in contributions to the Flora of Monmouthshire in the *Journal of Botany* for 1902, 1903 and 1904.

15

INTRODUCTION

Dr. George Claridge Druce (1850-1932) recorded in the *Journal of Botany*, vol. 46, 1908, a list of 17 species which he noted on a visit to the county.

The Rev. H. J. Riddelsdell (1866-1941) reported in the *Journal of Botany*, vol. 46, 1908, the discovery of *Ranunculus lingua* in a locality in which it appeared to be dying out. He kept the locality secret; it has not therefore been possible to ascertain whether or not the plant has become extinct. He recorded *Pyrus cordata* from the Dixton district, in the *Journal of Botany*, vol. 49, 1911.

Samuel Hamilton, B.A., M.B., of Newport, who was medical officer of health for the Marshfield district of the Newport Union, published in 1909 *The Flora of Monmouthshire*. This is a slim octavo volume of 81 pages. It is ostensibly a flora for the entire county, but most of the records are personal ones and relate chiefly to the Newport area. He gives 626 flowering plants and 23 ferns and fern allies. The small geographical handbook, *Monmouthshire*, by G. W. and J. H. Wade published in 1909 contains "Notes on the Flora" contributed by W. A. Shoolbred.

The last and one of the most important of contributions to the flora of Monmouthshire so far as the flowering plants are concerned was made in 1920 by W. A. Shoolbred when he published his *Flora of Chepstow*. This covered a radius of about 12 miles around Chepstow and includes approximately equal parts of Monmouthshire and West Gloucestershire. Many records from previously published sources were included but the greater proportion were made by Shoolbred himself and represented the accumulated observations of about 50 years.

The late S. G. Charles who resided in Monmouth made an invaluable contribution to the knowledge of the flora of Monmouthshire by his thorough exploration of the north-eastern part of the county between the years 1924-51. Most of the records he made are substantiated by herbarium specimens which he donated to the National Museum of Wales.

INTRODUCTION
THE BOTANICAL DISTRICTS
District 1

This district is the northern part of the county bounded on the south by a line drawn from the county boundary with Breconshire west of Abergavenny and following the Monmouthshire and Brecon Canal south-eastwards to Llanover and thence eastwards along the River Usk to Clytha Park. From that point the boundary follows the road to Monmouth via Raglan, with a short diversion north of the latter town, thence south-east to Redbrook.

Except for a small area of Carboniferous Limestone in the extreme eastern corner, the district is almost entirely on the Old Red Sandstone. The mean annual rainfall varies from 30 ins. in the east and 40 ins. in the south-west to 60 ins. in the north-west. The Honddu Valley, sometimes referred to as the Llanthony Valley, projects into the Black Mountains. Here the mountain cliffs of Tarren-yr-Esgob cross the boundary between Monmouthshire and Breconshire. A little to the south of these cliffs is Chwarel y Fan which at 2,228 ft. is the highest point in the county.

Amongst the plants recorded from the Honddu Valley are *Asplenium viride, Chrysosplenium alternifolium, Cystopteris fragilis, Eriophorum latifolium, Festuca altissima, Geranium sylvaticum, Gymnocarpium dryopteris, G. robertianum, Lathraea squamaria, Luzula forsteri, Lycopodium clavatum, L. selago, Meconopsis cambrica, Pinguicula vulgaris, Platanthera bifolia, Prunus padus, Saxifraga hypnoides, Sorbus porrigentiformis, Spiranthes spiralis, Thelypteris phegopteris, Vaccinium vitis-idaea, Veronica serpyllifolia var. humifusa* and *Viola lutea*.

On the limestone north of Monmouth is Lady Park Wood, forming part of the Highmeadow Woods, floristically one of the richest areas in the county. It is dominated by *Fagus sylvatica* with *Quercus robur* and *Fraxinus excelsior. Tilia cordata* is locally abundant and *Sorbus torminalis, S. aria* and *Tilia platyphyllos* occur on the rocky escarpments.

Other plants worthy of special mention are *Acinos arvensis, Atropa belladonna, Brachypodium pinnatum, Campanula latifolia, C. trachelium, Cardamine impatiens, Carex digitata, C. strigosa, Convallaria majalis, Cystopteris fragilis, Daphne laureola, Dipsacus pilosus, Festuca altissima, Gymnocarpium robertianum, Hordelymus europaeus, Hypericum montanum, Luzula forsteri, Melica nutans, Neottia nidus-avis, Ophrys insectifera* and *Rosa micrantha*.

Away from the neighbourhood of Monmouth and the region of Mynydd Pen-y-fal (Sugar Loaf), the woods are small, scattered and mainly of ash-oak.

The following plants which are very rare in the county are to be found in District 1:—*Bidens cernua, Oenanthe aquatica* and *Potentilla palustris* at Llanfoist; *Campanula patula, Polygonum mite* and *Stellaria nemorum* about Monmouth; *Alopecurus aequalis* at Penpergwm; *Dactylorhiza*

17

INTRODUCTION

incarnata at Pont-y-spig and *Pseudorchis albida* in the Grwyne Fawr Valley.

District 2

This district on the western side of the county includes the coalfield. It is bounded on the north by Breconshire and on the west by Glamorgan. On the east the boundary follows the Monmouthshire and Brecon Canal from Llanover to Pontypool and a line drawn from Pontypool through Risca to Machen.

The district is almost entirely on the coal measures but a narrow belt of Millstone Grit and Carboniferous Limestone lies along the eastern border with a larger area on Millstone Grit in the north-east. There is a small area of Carboniferous Limestone in the north-west. The mean annual rainfall increases gradually from about 45 ins. in the south to 60 ins. in the north. The district is mountainous and is cut by the narrow valleys of several rivers which flow southwards more or less parallel to one another. The valley bottoms are highly industrialised.

The hillsides up to 1,000 - 1,200 ft., once clothed with sessile oakwood and birch, are generally covered with bracken interspersed with *Agrostis-Nardus* grassland. On well-drained ground above the bracken, *Nardus stricta* becomes dominant, frequently associated with *Agrostis tenuis*, *Calluna vulgaris*, *Deschampsia flexuosa*, *Festuca rubra* and *Vaccinium myrtillus*. The latter often becomes dominant on the hill summits. On less dry slopes and mountain tops, *Molinia caerulea* usually forms a more or less level sward with *Juncus squarrosus*. Under wet conditions, upland moor dominated by *Molinia caerulea* occurs, with boggy patches here and there. True bogs where peat has developed are but poorly represented. They are small and confined to the headwaters of the Sirhowy and Ebbw rivers. Limestone pasture occurs above Trefil with *Aira caryophyllea*, *Briza media*, *Carex flacca*, *Carlina vulgaris*, *Circium acaule*, *Festuca rubra* and *Thymus drucei*. *Carex lepidocarpa* grows by the stream running through this area.

Rocks of the Pennant Grit series are exposed in a few places, forming low cliffs. Such cliffs are very barren and the rock ledges usually carry vegetation typical of the surrounding moorland or grass heath.

Much of the natural woodland which remained after the first World War has been felled and planted with conifers so that only fragments of the sessile oakwood remain. Relics of native beechwood are frequent on the eastern side of the district. Of particular interest is the beechwood on the limestone escarpment between Blaenavon and Llanelen, where *Circaea intermedia* was first recorded by Purton in 1817 and is still abundant. On the talus above this wood, *Gymnocarpium robertianum* is also abundant.

The following species rare in the county occur in District 2:—
Rhynchosinapis cheiranthus, abundant on the disused railway track below

18

INTRODUCTION

Mynydd Dimlaith, *Vicia orobus*, Abercarn and Crumlin; *Lathyrus nissolia*, Machen; *Potentilla palustris*, Blorenge; *Vaccinium oxycoccus* and *V. vitis-idaea*, near Blaenavon; *Littorella uniflora*, Pen-y-fan Pond; *Antennaria dioica*, Blorenge and Twmbarlwm, Risca and *Lagarosiphon major*, established in abundance in a quarry pool near Gelligroes.

District 3

This is the central region bounded by District 2 and the River Rymney from Machen southwards to the Rumney Bridge on the west, by District 1 on the north, and on the east by a line drawn from a point one mile east of Raglan and following the track of the old railway line to Llandenny, thence along the Olway Brook to its junction with the River Usk with which it coincides as far as Aber-nant, thence south to Cat's Ash and along the road to Christchurch and south to the railway. The southern boundary follows the main road from Cardiff to Newport and the railway from Newport to the Severn Bridge.

The district is for the most part on the red marls, but also includes the oval-shaped area of Upper Silurian rocks west of Usk. The mean annual rainfall is about 40 ins. The district is almost entirely agricultural, chiefly grazing, with a large number of small woods scattered throughout. The woods are ash-oak with *Quercus robur* and *Fraxinus excelsior* co-dominant. Associated trees are *Acer campestris*, *Betula pendula*, *Fagus sylvatica* and *Sorbus aucuparia*. The ground flora is rich in species. Alder-willow wood occurs in the wetter parts of the oak-ash wood and along streams.

Marsh vegetation is represented for the most part along streams and on the sites of former ponds, but a particularly interesting marsh is in the Henllys Parish. Here may be found *Carex paniculata*, *C. pulicaris*, *Cirsium dissectum*, *Epipactis helleborine*, *E. palustris*, *Eriophorum latifolium*, *Drosera rotundifolia*, *Galium uliginosum*, *Gymnadenia conopsea*, *Listera ovata*, *Menyanthes trifoliata*, *Pinguicula vulgaris*, *Polygonum bistorta* and *Valeriana dioica*.

The abundance of *Impatiens glandulifera* has long attracted attention along the banks of the river in the neighbourhood of Usk. *Blackstonia perfoliata*, *Carlina vulgaris*, *Cynoglossum officinale* and *Picris hieracioides* are to be found on the Wenlock Limestone west of Usk.

District 4

The eastern area situated between District 3 and the River Wye, bounded by District 1 on the north and on the south by the railway line from Liswerry to the Severn Bridge. It is mainly on the Old Red Sandstone but includes a large area of Carboniferous Limestone between Magor and Chepstow. To the south of the limestone are beds of Trias and near Llanwern is a cap of Lias resting on a thin bed of Rhaetic. In the region of Wentwood and between Chepstow and Monmouth the

19

INTRODUCTION

district is hilly and extensively wooded; elsewhere it is undulating and mainly pasture and hay meadow with numerous patches of grass-heath. The hills range from 400 to 1,003 ft.

The mean annual rainfall in the hilly area is 40 to 50 ins., and on the lower ground east of Usk and Raglan it is 40 ins.

Most of the woods at Wentwood and in the Wye Valley have been felled and replanted by the Forestry Commission or coppiced and interplanted with conifers. Oak has been planted as a pure crop on the Old Red Sandstone, whilst ash and beech have been planted on the limestone. Conifers have been planted on the relatively infertile soils which occur on the more exposed parts. Woods of mixed conifer and broad-leaved trees have also been planted.

Native woodland on the limestone still survives at Wynd Cliff, Black Cliff, Piercefield Park and Liveoaks. The dominant trees are *Fagus sylvatica, Fraxinus excelsior, Quercus robur, Q. petraea, Taxus baccata* and *Tilia cordata. Betula pubescens, Prunus avium, Sorbus* spp. and *Ulmus glabra* are frequent. Shrubs and climbers include *Clematis vitalba, Daphne laureola, Ligustrum vulgare, Rubia peregrina, Rubus saxatilis* and *Viburnum lantana.* Of herbaceous plants recorded from these woods are:— *Atropa belladonna, Carex digitata, C. montana, C. strigosa, Cardamine impatiens, Cephalanthera longifolia, Chrysosplenium alternifolium, Convallaria majalis, Euphorbia serrulata, Festuca altissima, Geranium pratense, G. sanguineum, Hypericum hirsutum, H. montanum, Luzula forsteri, Melica montana, Monotropa hypophegea, Neottia nidus-avis, Paris quadrifolia, Polygonatum odoratum, Pyrola minor, P. secunda, Scabiosa columbaria* and *Vicia sylvatica.* In a fragment of natural wood near Chepstow, *Carpinus betulus* occupies an outlier from its main area of distribution in Southern England.

Apart from Wentwood, the woods on the Old Red Sandstone are very small and scattered and are similar to the ash-oak woods on the Red Marl of District 3.

Limestone grassland is represented in a few places, notably in the neighbourhood of Rogiet and Carrow Hill where the following occur:— *Anacamptis pyramidalis, Anthyllis vulneraria, Astragalus glycyphyllos, Blackstonia perfoliata, Bromus erectus, Calamagrostis epigejos, Centaurea scabiosa, Cirsium acaule, C. eriophorum, Geranium columbinum, Helianthemum nummularium, Helictotrichon pubescens, Koeleria gracilis, Lithospermum officinale, Ophrys apifera, Plantago media, Poterium sanguisorba* and *Spiranthes spiralis.*

Trelleck Bog, an area of lowland moor or Molinietum, about 1 mile south-east of Trelleck, is on the site of a glacial lake. The peat is shallow and the vegetation is almost completely dominated by *Molinia caerulea.* The bog is slowly drying out due to its water being used to feed an emergency water tank. Gorse and bracken have invaded the eastern part. Saplings of *Sorbus aucuparia, Salix atrocinerea* and *S. aurita* are increasing in number and self-sown *Pinus sylvatica* also occurs. *Eriophorum angusti-*

PLATE I River Wye above Chepstow with cliff of Carboniferous Limestone and steep slopes with deciduous trees and yew. *Photo:* J. G. C. Anderson

PLATE II   Rhymney river meandering across estuarine clay flats east of Cardiff.   *Photo:* H. Tennant Ltd.

INTRODUCTION

folium, E. vaginatum and *Narthecium ossifragum* are abundant. *Drosera rotundifolia, Eriophorum latifolium, Hypericum elodes, Pinguicula vulgaris, Platanthera bifolia, Rhynchospora alba* and *Vaccinium oxycoccus* occur or have been recorded. There is a smaller and botanically less rich *Molinietum* at Whitelye Common.

In addition to those mentioned above the following noteworthy species have been recorded from District 4:— *Littorella uniflora* recently recorded from Wentwood Reservoir; *Lithospermum purpurocaeruleum,* near Carrow Hill; *Orobanche purpurea,* Mathern; *Polygonum mite,* by the River Wye, Llandogo and Whitebrook; *Ranunculus lingua,* near Trelleck; *Salvia pratensis,* Rogiet; *Stellaria nemorum,* Tintern and Llandogo.

District 5

This district is the coastal strip bounded on the north by the main road from Cardiff to Newport and by the railway line from Newport to the Severn Bridge. The land is low, mainly below 25 ft. O.D. and on estuarine and fluviatile clays reclaimed from the Bristol Channel. It is protected by sea walls and drained by an intricate network of deep ditches called reens. Below the sea walls are the wharflands covered only at high spring tides. They bear a close textured turf, heavily grazed by sheep and cattle, in which *Puccinellia maritima* is the dominant grass, associated with *Armeria maritima, Triglochin maritimum, Parapholis strigosa, Hordeum marinum, Alopecurus bulbosus, Carex distans,* etc. *Puccinellia distans* and *P. rupestris* colonize bare patches resulting from the removal of the turf. On the mud flats, separated from the wharfland by an earth cliff, *Salicornia europaea, S. dolichostachya* and *S. ramosissima* occur but are being overwhelmed by the rapidly increasing colonies of *Spartina anglica*.

The reens have a rich aquatic and marginal flora, which is unfortunately being threatened by the use of herbicides. Among the aquatic species are:— *Azolla filiculoides, Butomus umbellatus, Callitriche obtusangula, C. intermedia, Ceratophyllum demersum, C. submersum, Hippuris vulgaris, Oenanthe aquatica, Potamogeton crispus, P. pectinatus, P. pusillus, P. trichoides, Myriophyllum verticillatum, Ranunculus baudotii, R. circinatus, Ruppia maritima, Sagittaria sagittifolia* and *Utricularia neglecta.* Marginal species include *Althaea officinalis, Impatiens capensis, Lysimachia vulgaris, Lythrum salicaria, Rorippa amphibia, Rumex hydrolapathum* and *R. palustris*.

Other species of outstanding interest in the district are:— *Carduus tenuiflorus, Caucalis nodosa, Sagina maritima* and *Trifolium ornithopodioides,* on the tops of the sea walls; *Baldellia ranunculoides* on marshy ground at Magor and Rumney; *Bupleurum tenuissimum* at Rumney and Magor Pill; *Oenanthe lachenalii* at Rumney and Nash, and *Trifolium squamosum* in pastures by the sea wall at Rumney.

The tiny Denny Island, about $2\frac{1}{2}$ miles south-east of Magor Pill is the home of *Lavatera arborea*.

21

INTRODUCTION
GEOLOGY OF MONMOUTHSHIRE
By D. Emlyn Evans

The area now occupied by Monmouthshire, in common with most parts of the earth's surface, has had a long and complicated history. It has often been submerged beneath open seas from which it has been elevated repeatedly to form new land, and other strangely diverse geographical conditions such as forest swamps, desert wastes and polar ice sheets have existed at various times in the geological past. Each has given rise to the formation of different rock and soil types and these in turn have influenced directly or indirectly the present distribution of plants in the county.

The rocks of Monmouthshire are predominantly of the sedimentary kind, having originated as sediments, either as gravel, sand, mud or limestone produced by the weathering and destruction of pre-existent rocks or by the accumulation of organic remains. Such deposits, after inconceivably long periods of time, have been cemented by natural processes to form solid rocks which in turn have been compressed often into great folds and thus elevated to form new land which has been subjected immediately to erosion by the elements. It must be remembered, however, that there are within the county boundaries two occurrences of an igneous rock known as monchiquite. These minute relics of ancient volcanic activity are found at Great House near Golden Hill and at Glen Court, Llanllywel.

In order to account for the relief, structure and distribution of various rock types in the county it is advisable to outline the main events in the geological history. It is usual in such an exercise to divide geological time, as is historical time by the historian, into major and minor divisions. The table summarises the stratigraphical succession as it applies to the British Isles as a whole. Rocks of the Cambrian, Ordovician and Silurian Systems were first studied in Wales and the Borderland and were named after Wales (Cambria) or after ancient British peoples who lived in North-west Wales (Ordovices) and along the Welsh marches (Silures). Neither the Cambrian rocks, in which the oldest unmistakeable fossils can be found, nor the Ordovician rocks occur at the surface in Monmouthshire; nor, for that matter, do the oldest rocks of all that can be seen at the earth's surface, the so-called Pre-Cambrian rocks.

The Silurian System is usually divided into three Series which, in ascending order, are the Llandovery, the Wenlock and the Ludlow. The oldest rocks exposed at the surface in the county are the shelly mudstones known as the Wenlock Shale. This division, some 400-500 ft. thick, was laid down in a sea which covered the Welsh borders as well as a large part possibly of southern England. Above comes the Wenlock Limestone consisting of 35-45 ft. of sediment containing abundant fossils of limestone-forming shelly creatures whose occurrence suggests that there was a temporary shallowing of the sea due to uplift of the whole region by earth movements. This was followed by another deeper water period during which some 800 ft. of thick mudstones of the lower part of the Ludlow Series were laid down. The succeeding deposits, 100 ft. or so in thickness, consist mainly of impure limestones usually known as Aymestry Limestone. This reflects a further uplift of land with the ensuing shallowing of the sea. The uppermost parts of the Ludlow beds consist of some 250 ft. of siltstones often of a calcareous nature and obviously deposited in a deltaic or estuarine environment.

The earth movements which occurred towards the end of the Silurian period continued with increasing magnitude into the succeeding Devonian period and led to the formation of mountain ranges now known as the Caledonian mountains in Scandinavian, Scottish and North Wales regions. In our area, however, the earliest of these movements were sufficient only to cause the recession of the Silurian sea and the formation of a large estuary or delta into which the inflowing rivers from the northern mountains brought enormous loads of gravels, sands and clays which were eventually consolidated to form the conglomerates, sandstones and marls respectively of the Devonian System.

The open sea which had receded away to the south, covered, during Devonian times, the area now occupied by South Devon and Cornwall. The rocks belonging to the System and which are now found at the surface in these areas were deposited under marine conditions, whilst those in the areas to the north, including Monmouthshire, were deposited under continental or estuarine conditions. These two main groups of rocks, whilst belonging to the same System, are said to be of different facies, the former being described as the Marine Devonian and the latter as the Old Red Sandstone facies, respectively, of the Devonian System.

The Silurian rocks of our area are accordingly followed upwards by rocks of the Old Red Sandstone. These are so called because they consist largely of red-coloured rocks and are older than the Coal Measures in *contra* distinction to the New Red Sandstone which overlies the Coal Measures.

The change from the marine conditions of Silurian times to the estuarine conditions of the Old Red Sandstone facies resulted first in the disappearance of the Silurian creatures and eventually provided the necessary environment for the development and evolution of fishes. By reference to the fossil fish the whole facies has been divided into lower, middle and upper series. In the Lower Old Red Sandstone, fossil specimens have been found of primitive jawless fishes known as ostracoderms which reached their acme in Lower Devonian times and which are comparable with the present day lampreys.

The basal beds of the Series consist of buff and yellow sandstones varying in thickness from 5 to 75 ft. and known as the Downton Castle Sandstone. They represent a continuation of the deltaic sedimentation

INTRODUCTION

that was commenced in late Silurian times so that there is no readily drawn dividing line between the Silurian and the Old Red Sandstone. In the south of the outcrop in the county, however, there occurs a poorly developed bed of sandstone some 3 in. thick containing fish remains and known as the Ludlow Bone Bed. The base of this bed provides a convenient boundary between the two systems.

Above comes the so-called Raglan Marl Group of some 1,100-2,000 ft. thick and consisting mainly of red marls with subordinate sandstones and concretionary limestones probably deposited in a low-lying and gradually subsiding delta plain. At the top of the group there occurs 100 ft. or so of marls with limestone nodules and nodular and massive concretionary limestones becoming more abundant upwards, the sequence terminating in the well-developed *'Psammosteus'* Limestone. The formation of these thick limestones marks a quiet period when most of the area was flooded by shallow, brackish water in which little or no sediment was deposited.

The succeeding St. Maughan's Group of some 1,500-2,000 ft. thickness and consisting primarily of interbedded marls with sandstones, conglomeratic cornstones, concretionary limestones and conglomerates, commences with coarse deposits which indicate a sudden uplift of the Caledonian mountains and the commencement of a period when fluviatile conditions became more fully established over the delta. Higher in the group the sediments become finer grained and concretionary limestones appear suggesting conditions of deposition similar to those which prevailed at the end of the Raglan Marl Group times. The upper 450 ft. of the group are characterised in the area south and south-west of Machen by the Llanishen Conglomerate. This is considered to have been derived from a land ridge which probably emerged during late St. Maughan's Group times along the site of the present Bristol Channel.

The next division is that of the Brownstone Group which consists primarily of reddish-brown sandstones 400-600 ft. thick. They denote renewed uplift of the Caledonian mountains from which sand was brought into the area by large rivers. This deposition was terminated by yet another major earth movement which attained great intensity to the north-west of our district, in Central and North Wales. It seems that throughout Middle Devonian times erosion took place in our area so that the Middle Old Red Sandstone is completely unrepresented. The earth movements, though slight in comparision to other areas, were sufficient to produce uplift and some folding. Considerable erosion took place before the region again sank beneath the waters of a large delta. The evidence for these conclusions is of particular interest. It has been found that the quartz-conglomerate which occurs at the base of the Upper Old Red Sandstone lies uncomformably upon the Lower Old Red Sandstone. It lies in places with little apparent break upon the Brownstones but in other

INTRODUCTION

places on the St. Maughan's Group or the Raglan Marl Group and even on the Silurian in the Tortworth area.

The Upper Old Red Sandstone consists of a laterally impersistent basal conglomerate which is overlain by alternations of variably coloured sandstones and silty mudstones which originated as sub-aerial deposits on the delta. The whole division ranges in thickness from some 250 to 400 ft.

The Old Red Sandstone was brought to a close and the succeeding Carboniferous System ushered in by subsidence of land which allowed for our area to be quietly inundated by a shallow shelf sea in which limestones and shelly mudstones were deposited. These rocks form the lowest litho-logical division of the Carboniferous Limestone and are known as the Lower Limestone Shale. They occur in the western and eastern regions of the county and have an average thickness of 120 ft.

In the Monmouthshire coalfield region all the divisions of the Car-boniferous Limestone above the Lower Limestone Shale are best described as the Main Limestone which consists of thick organic limestones, most of which were changed from pure calcareous deposits into hard, grey, brown-weathering, thickly bedded dolomitic limestones during or soon after deposition. This means that except for crinoid ossicles few fossils have survived dolomitization, so that precise palaeontological correl-ation between different areas is not possible. The thickness of the Main Limestone decreases from about 1,600 ft. in the Machen area to a little over 500 ft. in the Llangattwg area. This, together with the fact that some of the higher beds are entirely absent in the latter area, suggests that the limestones were deposited in a marine trough bounded on the north and north-west by land and on the south and south-east by a submarine ridge which extended south-westwards through the area now occupied by the town of Usk. This ridge, which separated the area of deposition of the limestones of the Monmouthshire coalfield from that of the Chepstow and Forest of Dean limestones, was being gradually uplifted until eventually in Millstone Grit times its continued elevation gave rise to the formation of land, so that no Millstone Grit deposits occur in the Forest of Dean coalfield.

In the Chepstow district the Lower Limestone Shale is succeeded by some 10 ft. of grey crinoidal limestone and above this are some 40 ft. of grey-white oolite. Both limestones are collectively known as the Crease Limestone. At the top of the oolite there is definite evidence of erosion which obviously resulted from uplift of the sea bed. The sub-sidence which followed this uplift led to the deposition in a shallow sea of the Whitehead Limestone, which consists mainly of pale grey calcitic mudstone together with dolomitic mudstone. The overlying Lower Dry-brook Sandstone which is just over 70 ft. thick in Chepstow, represents another marked change in sedimentation. In places at the base there occur conglomerates, but the formation consists mainly of reddish-pink

INTRODUCTION

or yellow sandstones composed of rounded quartz grains. The next division is that of the Drybrook Limestone which represents the deposition of calcareous deposits in a sea which developed in the Chepstow-Forest of Dean area. Deeper water obviously lay to the south so that the limestone thickens in that direction, a process which takes place at the expense of the underlying sandstones, there being a lateral passage from limestone into sandstone. This limestone formation consists of some 400 ft. of oolites containing abundant fossils of *Composita* and *Lithostrotion*.

The Carboniferous Limestone deposition was brought to a close by a major earth movement which caused uplift and rock folding in our area. As a result much of the eastern part of the region suffered great erosion. Following the uplift there came one of the longest periods of continued but oscillatory subsidence of land that has occurred in the whole of British geological history. It was during the earliest part of this subsidence that the next series of rocks, the Millstone Grit, was deposited, but only in the western part of our area. It is not surprising to find that, as a result of the changed geographical conditions under which they were laid down, these rocks differ greatly from the underlying limestones. They consist mainly of two groups, a lower variable group of sandstones with conglomerates, and an upper shale division, all of which were derived from the north and carried southwards by the rivers where they were deposited in large estuaries.

Generally speaking the Millstone Grit deposits become coarser when traced to the north and it is assumed that the northerly contemporary shoreline lay but a little distance beyond the present outcrop and deep waters extended over the Bristol Channel area. The greatest thickness of the Series amounting to some 2,000 ft. is found in the Gower. This thickness is reduced to 1,000 ft. in the Llandebie area but when traced to Pontypool the whole Series is reduced to a little over 100 ft. In the Chepstow and Forest of Dean areas no Millstone Grit deposits are recorded between the Carboniferous Limestone and overlying Coal Measures. From the foregoing it can be deduced that each succeeding bed of the Series was deposited a little further east than the underlying bed as the land which lay in this direction was being gradually submerged. This process extended throughout Millstone Grit times and well into the following Coal Measures times because the Coal Measures of the Forest of Dean can be correlated only with the very highest measures represented in the South Wales coalfield.

The corals and brachiopods of the Carboniferous Limestone are replaced in the Millstone Grit by animals adapted to life in a shallow water, muddy, marine or brackish environment. These are chiefly lamellibranchs and goniatites, the latter related to the present-day *Nautilus*. The present classification of the Millstone Grit is based on its sequence of goniatite faunas and the divisions, depending mainly on the work of

INTRODUCTION

Bisat in northern England, are in ascending order as follows:— Pendleian Stage or Lower *Eumorphoceras*, E_1; Arnsbergian Stage or Upper *Eumorphoceras*, E_2; Chokierian Stage or Lower *Homoceras*, H_1; Alportian Stage or Upper *Homoceras*, H_2; Kinderscoutian Stage or Lower *Reticuloceras*, R_1; Marsdenian Stage or Upper *Reticuloceras*, R_2; Yeadonian Stage or Lower *Gastrioceras*, G_1.

Although the Millstone Grit is poorly exposed along the southern and eastern edges of the South Wales coalfield, faunal evidence has been obtained in our area to show that only beds of G_1 age are represented. In addition to the absence of the lower strata there is also, as one moves to the east, a remarkable thinning of all beds. This feature is obviously due to the effects of deposition against a westerly inclined shoreline. The occasional bands of truly marine shale which contain the distinctive goniatites are often represented in the eastern areas by extremely thin marine sediments containing no recognisable goniatites or any other fossils. Along the North Crop from Rhymney Bridge to the Blorenge area the same features occur but there is faunal evidence in this instance that beds of R_1, R_2 and G_1 age occur in Rhymney whilst the lower beds in the Blorenge area seem to be of R_2 age.

The sandstones and conglomerates (which in places include a few intercalated mudstones) at the base of the Millstone Grit vary in thickness from 0 to 100 ft., while the overlying mudstone division generally varies from 70 to 175 ft.

For many years the traditional upper limit of the Millstone Grit in the eastern part of the South Wales coalfield had been drawn at the top of a sandstone known as the Farewell Rock and which was the first thick sandstone below the lowest economically important seam of coal in the area, the Old Coal or Hard Vein. It was usual to divide the Millstone Grit into three logical lithological units, an Upper Farewell Rock, a Middle Shales and a Basal Grit division. In 1957, however, an international recommendation was accepted and the upper limit of the Millstone Grit was drawn at the base of a marine bed which contains the goniatite *Gastrioceras subcrenatum* C. Schmidt and which occurs below the Farewell Rock. This meant that the Farewell Rock and some shales below were transferred into the Lower Coal Measures division.

The dominantly marine sediments of the Millstone Grit were followed by Coal Measures whose total thickness varies from a little under 1,500 to a little over 4,000 ft. and which were deposited under fresh water or estuarine conditions with occasional incursions by the sea. The oscillatory subsidence which had commenced during the period of deposition of the Millstone Grit, that is in Namurian times, continued into and persisted through Coal Measures times. At regular intervals, most frequently in the earlier stages, the area was covered by extensive forest growth. Each forest period was terminated by subsidence of land which caused flooding

INTRODUCTION

of the area often by the sea. This resulted in the decay of the forest vegetation which, on burial by sediments accumulated by incoming rivers, was converted into coal. Continued deposition of sediments gave rise to the infilling of the estuarine lagoons so that another growth of forest vegetation was possible. The ideal cyclic pattern associated with such deposition consists of the following ascending sequence:—mudstone with marine fossils, mudstone with non-marine fossils, generally unfossiliferous mudstone with plants towards the top passing up into sandstone, sandstone passing up into mudstone, seatearth and coal.

In the Coal Measures of our area there are over thirty such cyclic units, although there are some in which there is no commercially important coal seam included. The most important single deposits for classification purposes are the marine bands. Reference has already been made to the *Gastrioceras subcrenatum* Band at the base of the Coal Measures, equivalents of which are found in all other British coalfields as well as those in north-west Europe. Two other equally important marine bands occur. The lower is characterised by the goniatite *Anthracoceras vanderbecki* (Ludwick) and known in South Wales as the Amman Marine Band. The upper is characterised by the goniatite *Anthracoceras cambriense* Bisat and known in South Wales as the Upper Cwmgorse Marine Band. The whole of the Coal Measures division is divided into lower, middle and upper sections, the base of the middle being drawn at the base of the Amman Marine Band and the base of the upper at the top of the Upper Cwmgorse Band, and not at the base as is more usual. This has been done because the horizon marks the close of a particular paralic type of deposition. No marine bands occur above so that the Upper Coal Measures can be described as including no marine bands.

The lower part of the Upper Coal Measures contains a number of beds of red mudstone and seatearth whose colouration is thought to be due to oxidation of sediment under tropical weathering. Above these red rocks the Upper Coal Measures are characterised by thick deposits of current-bedded sandstone. This is the so-called Pennant Sandstone which occupies large tracts of the South Wales coalfield plateau surface and which until 1957 played such an important part in the classification of the Measures. The old threefold divisions were known as the Lower Coal Series, the Middle or Pennant Sandstone Series and the Upper Coal Series. It has been shown that the base of the sandstone occupies in the eastern part of the South Wales coalfield a much higher position in the geological column than it does further west, so that the term "Pennant" should not henceforth be used in a stratigraphical sense but should refer only to the rock type.

At the end of Coal Measures times there occurred a major climatic change. Desert conditions prevailed and during the early part of this

INTRODUCTION

new period major earth movements occurred associated with the Armorican Mountain Building movement. Rocks of the Carboniferous and earlier formations of the county were folded to form synclines or downfolds separated by anticlines or upfolds, all aligned in a general east-west direction. From a recent age determination of the Golden Hill monchiquite it has been established that this was the time at which the intrusion occurred. During the remainder of the period now known as the Trias, the mountainous crests of the anticlines were largely eroded away so that the Carboniferous strata were preserved only in the synclines, whilst the older strata were often revealed in the worn cores of the anticlines. This was the time when most of the rocks that once covered the centre of the county were removed.

In late Triassic times the Bristol Channel area was occupied by an extensive salt lake into which some of the weathered material from the anticlinal mountains around was deposited. The basal member, the Keuper dolomitic conglomerate, represents scree and downwash and is often about 20 ft. thick. Away from high ground the conglomerate grades through fine breccia into Keuper Red Marl which was originally a muddy lake deposit and which has a thickness of over 300 ft. This passes upwards into some 20 ft. of Tea Green Marl whose colour may indicate the approach of more humid conditions towards the end of the period.

With further subsidence of land the salt lake was flooded by the open seas in which rocks of the lower divisions of the succeeding Jurassic System, the Rhaetic and the Lias, were deposited upon the unfossiliferous Triassic rocks below and are now preserved in the county as outliers capping the Triassic outcrops from Bishton to Chepstow. The Rhaetic is divisible in ascending order into the Westbury and Cotham Beds, the former being about 16 ft. thick and consisting mainly of shales, whilst the latter are 8 to 9 ft. thick and consist mainly of calcareous mudstones. The Lower Lias in turn is divisible into some 80 ft. of Blue Lias limestone below and 120 ft. of so-called Blue Lias clay above. Most of these formations contain, amongst the marine fossils, reptilian remains and the Lias contains some of the earliest ammonites.

The 180 million years which elapsed between the deposition of the Lias rocks and the beginning of the Pleistocene period are not represented by any deposits in South Wales and so any attempt to reconstruct the geological history of this span of time must be largely speculative. It can be assumed that the British area as a whole was uplifted out of the Chalk Sea and that our present day rivers first flowed upon the then newly emerged land surface some 65 million years ago. It appears likely too that they were able to remove all the Cretaceous and most of the Jurassic rocks and were thus lowered or superimposed upon the folded rocks below. They were able also to produce at least one fairly flat peneplain, remnants of which now exist as the flat-topped bare hills between the coalfield valleys.

THE STRATIGRAPHICAL SUCCESSION

PERIOD OR SYSTEM	AGES IN MILLIONS OF YEARS		REPRESENTED IN MONMOUTHSHIRE
RECENT			Peat, alluvium, estuarine mud, landslips. 'Head'.
	—.012—		
PLEISTOCENE			Glacial deposits such as sand, gravel and boulder clay. 'Head'.
	—2—		
PLIOCENE MIOCENE OLIGOCENE EOCENE			
	—65—		
CRETACEOUS			
	—136—		
JURASSIC	LOWER LIAS		Lower Lias Clay—mainly mudstones with some limestones. Blue Lias—limestone beds with mudstone partings.
	RHAETIC		Cotham Beds—calcareous mudstones. Westbury Beds—shales and mudstones.
	—190–195—		
TRIASSIC	KEUPER		Tea Green Marl—mudstones. Keuper Marl—red mudstones. Dolomitic Conglomerate—breccias and/or dolomitic limestone.
	COAL MEASURES		Upper or Pennant Measures—mainly sandstones, some mudstones. Middle Measures—mudstones, some sandstones and many coals. Lower Measures—mudstones, some sandstones and many coals.
	—280—		
	MILLSTONE GRIT		Yeadonian Stage or G_1 zone / Mardenian Stage or R_2 zone } mainly shales. Kinderscoutian Stage or R_1 zone—predominantly sandstones and grits.
CARBONIFEROUS	—310–315—		

Age (Ma)	System	Series / Group	Newport District	Faunal Zones	Chepstow District
325		CARBONIFEROUS LIMESTONE SERIES	Main Limestone	Seminula Zone	Drybrook Sandstone
				Upper Caninia Zone	Whitehead Limestone
				Lower Caninia Zone	Crease Limestone
				Zaphrentis Zone	Lower Dolomite
			Lower Limestone Shales	Cleistopora Zone	Lower Limestone Shales
345	*DEVONIAN*	UPPER OLD RED SANDSTONE	Sandstones and marls above. Quartz conglomerates and grits below.		
		LOWER OLD RED SANDSTONE	Brownstone Group—mainly reddish brown sandstones. St. Maughan's Group—interbedded marls and sandstones. Raglan Marl Group—mainly red marls. Downton Castle Sandstone—grey and buff sandstones.		
395	*SILURIAN*	LUDLOW	Whitecliffe Beds. Leintwardine Beds. } mainly siltstones. Bringewood Beds—calcareous siltstones and limestones. Elton Beds—mainly silty mudstones.		
		WENLOCK	Wenlock Limestone—limestones. Wenlock Shale—mainly mudstones.		
430–440					
500	ORDOVICIAN				
570	CAMBRIAN				
4600	PRE-CAMBRIAN				

The Systems represented in Monmouthshire are printed in italic type.

INTRODUCTION

There followed a period of prolonged and staggered lowering of the sea level in relation to the land. At one stage it stood some 600 ft. higher than the present level and sea cliffs were formed in the southern edge of the coalfield plateau mainly along the steeply inclined beds of Pennant Sandstone. The sea level continued to fall at sporadic intervals between which the waves were able to cut gently inclined and stepped surfaces across the Vale of Glamorgan at average heights of some 400, 300 and 200 ft. above the present sea level. The rivers that flowed across the coalfield throughout this period were eroding their valleys to lower and lower depths as the level of the sea into which they flowed was continuing to fall. There is every possibility that the floors of the valleys were excavated in rock to a depth below the present position of the rivers. It is a well-known fact that the rivers of South Wales flow, in their lower reaches, upon clay and gravel which lie upon the buried valley floors.

During this long period of river erosion, regions that had been weakened by folding and faulting were quickly exploited by streams that were able to deepen their valleys more quickly than others and were able accordingly to capture or divert into their own valleys sections of neighbouring rivers. This process known as "river capture" has given rise, amongst other features, to the occurrence of abandoned valley sections or misfit streams. One of the best examples of the former landforms is to be seen in the Bettws gap immediately east of the Sugar Loaf. This was at one time occupied by the Grwyne Fawr stream until it was captured, as a result of headward erosion eastwards, by a tributary of the Grwyne Fechan which exploited the faulted zone, which is an eastward extension of the Vale of Neath disturbance and which extends from Penderyn to Dolygaer to Llangynidr through Cwm Coed y Cerrig, to Pandy and beyond. Another good example of an abandoned river valley is the Hafodyrynys gorge which leads westwards from Pontypool. This was probably the course of the Afon Lwyd stream which flowed westwards to join the Ebbw at Crumlin until it was captured by a stream flowing down the edge of the coalfield in the Pontypool area to join the Usk at Caerleon. It is probable too that the Afon Lwyd previously had an extension much further north on to the Mynydd Llangattwg plateau until it was captured by the Clydach stream as it extended its course westwards into the rim of the coalfield. It is probable, furthermore, that part of this beheaded section gradually descended into the Carboniferous Limestone that lies beneath the surface in this area and that this underground stream, by solutional weathering, formed eventually the extensive Agen Allwedd cave system, water from which drains underground into the bed of the Clydach.

Some two-and-a-half to two million years ago there commenced a period now known as the Pleistocene during which an alternating sequence of colder and warmer climates occurred. It has been popularly known as the Ice Age and is now considered to have terminated about ten thousand

INTRODUCTION

years ago when the so-called Recent period commenced. Both the Recent and Pleistocene periods are regarded as subdivisions of the Quaternary Era.

In Britain extensive glaciation occurred only in the Middle and Upper Pleistocene. South Wales was probably affected by at least three major ice advances, termed by Wells as the First Welsh Glaciation, the Second Welsh Glaciation and, lastly, the Welsh Re-advance or Little Welsh Glaciation. Although the general topography of our region was determined in pre-glacial times, the land surface was modified during glacial and inter-glacial times by erosion of rock surfaces and by the deposition of glacially derived material which is known as drift. The V-shaped cross sections of many of the pre-existent river worn valleys were altered by glaciers and became U-shaped. The resulting rock debris gave rise to two kinds of deposit. The unsorted drift which consists of angular and rounded boulders normally in a sandy, clayey matrix is known as boulder clay and is usually found in the upper regions of the glaciated valleys. In the lower parts of the valleys melt-water flowing beneath, upon or issuing from the ice has re-sorted the boulder clay to produce deposits of gravel and sand often at some distance from the furthest extension of the ice.

The interpretation of the glacial deposits in Monmouthshire and their relationship to the individual glaciations is difficult because the valley glaciers of each advance disturbed and possibly removed the deposits of the previous glaciation. The glacial deposits of the county are believed to be mainly due to the final ice advance, although it is still not known whether the high level drift and blocks of transported rock now found on the crests of the watersheds date from this period or are due to an earlier glaciation.

The extent of the last glaciation is marked by thick morainic deposits which extend from Machen to Newport and thence northwards up the Ebbw Valley to Brynmawr, down the Clydach Valley eastwards to the Usk, southwards to Mamhilad, eastwards to Raglan, north-westwards to the Abergavenny district and finally to Llanvihangel Crucorney.

A particularly interesting result of glaciation is the manner in which certain accumulations of drift were deposited at crucial points in some valleys, thereby causing the post-glacial rivers to be diverted into new channels. Ice which "over-spilled" from the Wye Valley in the Llyswen-Glasebury-Hay region through the Cnapiau coll at the head of the Honddu Valley moved down the Vale of Ewyas and left a large moraine at Llan-vihangel Crucorney. This barrier prevented the Monnow and Honddu rivers from following their original combined course through to the Usk at Abergavenny. The water that must have been impounded behind the moraine found an outlet along the present river route via the areas now occupied by Grosmont and Skenfrith to join the Wye at the point at which the town of Monmouth now stands.

INTRODUCTION

Towards the end of the Pleistocene period climatic conditions were still sufficiently cold for the precipitation of appreciable winter snow. It was under these so-called periglacial conditions that the snow-formed-cwms originated in some of the coalfield valleys, in the Blorenge region and also along the western sides of the Vale of Ewyas. It was at this time too that the process of landsliding was initiated and this has persisted to the present day especially in the coalfield valleys. Spectacular landslides are also to be seen on the Skirrid and along the eastern slopes of the Vale of Ewyas where steep sandstone scarps are underlain by impermeable marls. Another result of periglacial conditions was the formation of a layer of clayey stone-laden soil which slid slowly down the valley slopes over the permanently frozen sub-soil. This so-called "head" deposit covers large tracts of the county.

At the end of the glacial episode rivers in the upper reaches of the principal valleys cut into the glacial deposits and formed the channels seen today. Along the lower reaches of the valleys the early post-glacial rivers flowed along channels deeper than those of the present day because sea level was at least 100 ft. lower than it is now. Since then there has been a gradual rise in sea level and the lower parts of the valleys have been infilled with alluvial deposits normally to the present day level. Evidence of other very recent events in the geological history of the county is provided by the estuarine mud, which covers the coastal flats of the Bristol Channel, and the small peat bogs found in localities of poor surface drainage either on the higher moorlands or in hollows between the lowland glacial deposits.

The above account is based entirely on the following references:—

Robertson, T. (1927). The Geology of the South Wales Coalfield. Part II. Abergavenny. 2nd Ed. *Mem.geol.Surv.U.K.*

Squirrell, H. C. and Downing, R. A. (1969). Geology of the South Wales Coalfield. Part I. The Country around Newport (Mon.). 3rd Ed. *Mem.geol.Surv.U.K.*

Welch, F. B. A. and Trotter, F. M. (1961). Geology of the country around Monmouth and Chepstow. *Mem.geol.Surv.U.K.*

BIBLIOGRAPHY

Babington, C. C. (1845). A visit to Tintern. *Phytologist*, first series, **2**, 8.

Babington, C. C. (1869). *The British Rubi*. London.

Babington, C. C. (1897). *Memorials, Journal and Botanical Correspondence*. Cambridge.

Ball, John. (1849). Contributions to the Flora of South Wales. *Bot. Gazette*, **1**, 107-109.

Bennett, A. (1903). *Salvia pratensis* in Monmouthshire. *J. Bot., Lond.*, **41**, 285.

Bladon, J. (1857). *Aconitum napellus, Phytologist*, second series, **2**, 72.

Bull, H. G. (1870). The Mistletoe-oak at Llangattock Lingoed. *Rep. Woolhope Club*, 1868-69.

Clark, J. H. (1853). *Cardiff and its Neighbourhood*. Usk.

Clark, J. H. (1856). *Usk and its Neighbourhood*. Usk.

Clark, J. H. (1868). *The Flora of Monmouthshire*. Usk.

Clark, J. H. (1868). *Sketches of Monmouthshire*. Usk.

Donovan, E. (1805.) *Descriptive Excursion through South Wales and Monmouthshire*, **1**, London.

Druce, G. C. (1908). Welsh Records. *J. Bot. Lond.*, **46**, 335-336.

Evans, F. (1875). Some of the present inhabitants of Raglan Castle. *Trans. Cardiff Nat. Soc.*, 1874, **6**, 55-61.

Gissing, T. W. (1853). Notes of a Botanical Excursion down the Wye. *Phytologist*, **4**, 1053-1055.

Hamilton, S. (1909). *The Flora of Monmouthshire*. Newport.

Hooker, W. J. (1842). *The British Flora*. London. Ed v (1850), Ed. vi, jointly with G. A. Walker-Arnott.

Hort, F. J. A. (1851). On *Euphorbia stricta* and *E. platyphylla. Bot. Gaz.*, **3**, 15-17.

Hort, F. J. A. (1852). Notes on the Occurrence of *Aconitum napellus* and *Orobanche coerulea* in Monmouthshire. *Phytologist*, **4**, 640.

Howarth, W. O. (1924). On the Occurrence and Distribution of *Festuca rubra* Hack. in Great Britain. *J. Linn. Soc., Lond.*, **46**, 313-331.

Hyde, H. A. & Wade, A. E. (1957). *Welsh Flowering Plants*. Ed. ii, Cardiff

Hyde, H. A. & Wade, A. E. (1962). *Welsh Ferns*. Ed. iv, Cardiff.

Hyde, H. A, Wade, A. E. & Harrison, S. G. (1969). *Welsh Ferns, Clubmosses, Quillworts and Horsetails*. Ed. v, Cardiff.

Lees, E. (1837). A Botanical Tour in Herefordshire, Monmouthshire and South Wales. *Naturalist*, **1**, 209, 260; **2**, 115, 204, 254, 295.

Lees, E. (1842). *The Botanical Looker-Out*. London. Ed. ii (1851).

Ley, A. (1890). The Botany of the Honddu and Grwyne Valleys. *Trans. Woolhope Nat. Field Club*, 1883-1885, 343-351.

Ley, A. (1894). Additions to the Herefordshire Flora. *Trans. Woolhope Nat. Field Club*, suppl.

Ley, A. (1898). Two new forms of *Hieracium. J. Bot., Lond.*, **36**, 6-7.

Ley, A. (1900). Some Welsh Hawkweeds. *J. Bot., Lond.*, **38**, 3-7.

Ley, A. (1907). British Roses of the mollis-tomentosa Group. *J. Bot., Lond.*, **45**, 200-210.

BIBLIOGRAPHY

Lowe, E. J. (1867, 1869). *Our Native Ferns*. London.

Manby, G. W. (1802). *A Historic and Picturesque Guide . . . through the Counties of Monmouth, Glamorgan and Brecknock*. Bristol.

Marshall, E. S. (1916). Notes on *Sorbus*. *J. Bot., Lond.*, **54**, 10-14.

Marshall, E. S. & Shoolbred, W. A. (1902). Gloucestershire and Monmouth Plants. *J. Bot., Lond.*, **40**. 263-264.

Marshall, E. S. & Shoolbred, W. A. (1903). Monmouthshire Plants. *J. Bot., Lond.*, **41**, 140.

Marshall, E. S. & Shoolbred, W. A. (1904). Monmouthshire Plants. *J. Bot., Lond.*, **42**, 157.

Mathews, W. (1879). A list of Midland County Plants, 1849-1884. *Trans. Worc. Nat. Club*, 1897-1899, 55-127.

Matthews, L. H. (1932). Denny Island. *Proc. Bristol Nat. Soc.* **7**, 371-378.

Moore, T. (1859). *The Nature Printed British Ferns*. London.

Monington, H. W. (1888), New County Records. *J. Bot., Lond.*, **26**, 376.

Newman, E. (1840, 1844). *A History of British Ferns and Allied Plants*. Ed. i, 1840, Ed. ii, 1844. London.

Parkinson, C. (1894). Plants in Western England. *Sci. Gossip*, n.s. 1, 107.

Parkinson, J. (1640), *Theatrum Botanicum*.

Philipson, W. R. (1937). A Revision of the British Species of the Genus *Agrostis* L. *J. Linn. Soc., Lond., (Botany)*, **51**, 73-151.

Pugsley, H. W. (1930). A Revision of the British Euphrasiae. *J. Linn. Soc., Lond.*, **48**, 467-544.

Pugsley, H. W. (1948). A Prodromus of the British Hieracia. *J. Linn. Soc., Lond.*, (Botany), **54**.

Purchas, W. H. & Ley, A. (1889). *The Flora of Herefordshire*. Hereford.

Purton, T. A. (1817-1821). *The Midland Flora*. London.

Ray, J. Itinerary iii, 1662. In *Memorials of John Ray* edited by E. Lankester, 1846. London.

Ray, J. (1670). *Catalogus Plantarum Angliae*. London. Ed. ii (1677).

Ray, J. (1690). *Synopsis Methodica Stirpium Britannicarum*. London. Ed. 2 (1696). Ed. 3 (1724).

Reader, H. P. (1886). New Records from Gloucester and Monmouth. *J. Bot., Lond.*, 368-370.

Riddelsdell, H. J. (1905). Lightfoot's visit to Wales in 1773. *J. Bot., Lond.*, **43**, 290-307.

Riddelsdell, H. J. (1908). Monmouth Plants. *J. Bot., Lond.*, **46**, 231.

Riddelsdell, H. J. (1911). *Pyrus cordata* in Monmouthshire. *J. Bot., Lond.*, **49**, 170.

BIBLIOGRAPHY

Sandwith, C. & N. Y. (1953). Additions to the Flora of Denny. *Proc. Bristol Nat. Soc.*, **28,** 313-314.

Shoolbred, W. A. (1894). Recent additions to the Flora of West Gloucester and Monmouth. *J. Bot., Lond.*, **32,** 263-271.

Shoolbred, W. A. (1894). West Gloucester and Monmouth Plants. *J. Bot., Lond.*, **32,** 311.

Shoolbred, W. A. (1896). Plants of West Gloucester and Monmouth. *J. Bot., Lond.*, **34,** 29-30.

Shoolbred, W. A. (1896). New Monmouthshire Brambles. *J. Bot., Lond.*, **34,** 480.

Shoolbred, W. A. (1898). West Gloucester and Monmouth Plants. *J. Bot., Lond.*, **36,** 32, 402.

Shoolbred, W. A. (1920). *The Flora of Chepstow.* London.

Southall, H. (1880). On the Botany of the neighbourhood of Ross and the lower portion of the Wye Valley. *Trans. Woolhope Nat. Club,* 1875, 121-124.

Turner, D. & Dillwyn, L. W. (1805). *The Botanist's Guide,* **2.** London.

Wade, G. W. & J. H. (1909) *Monmouthshire.* London. Notes on the flora of Monmouthshire by W. A. Shoolbred.

Walford, T. (1818). *The Scientific Tourist.* London.

Warner, R. (1798). *A Walk through Wales in* 1797. Bath.

Watson, H. C. (1835). *The New Botanist's Guide,* **1,** 215. London. (1837). **2,** 629 and appendix. London.

Watson, H. C. (1873-74). *Topographical Botany.* London. (1883), Ed. ii.

Welch, B. (1950). Flora of Denny Island. *Proc. Bristol Nat. Soc.*, **28,** 22-23.

White, J. (1886). *Guide to the Town and Neighbourhood of Abergavenny.* Abridged edition. 74-76.

Whitwell, W. (1891). Monmouth County Records. *J. Bot., Lond.*, **29,** 308.

Withering, W. (1801). *Botanical Arrangement of British Plants.* Ed. iv, London.

Woods, J. (1850). Botanical Notes, the result of a visit to Glamorgan and Monmouthshire . . . *Phytologist,* **3,** 1053-1061.

37

ANNOTATED FLORAS AND MANUSCRIPTS

Guile, D. P. M. (1965). *The Vegetation of the Brecon Beacon National Park*. Ph.D. thesis. Copy in the National Museum of Wales Library.

Hudson, G. (1762). *Flora Anglica*, London. Copy in the Library of the British Museum (Natural History) annotated by J. Banks.

Mavor, William. *The Botanical Pocket Book for* 1800. Copy in the possession of J. S. L. Gilmour annotated by an unknown botanist in 1806 and 1807.

Ray, J. (1724). *Synopsis Methodica Stirpium Britannicarum*, London, Ed. iii. Copy in the Department of Botany, Oxford annotated by J. Lightfoot and J. Hill.

Shoolbred, W. A. MS. lists and notes relating to the Flora of the Chepstow district in the National Museum of Wales Library.

Shoolbred, W. A. (1920). *The Flora of Chepstow*, London. Copy in the National Museum of Wales Library annotated by the author.

Turner, D. & Dillwyn, L. W. (1805). *The Botanist's Guide*, London. Copy in the Library of the British Museum (Natural History) annotated by E. Forster.

LIST OF RECORDERS

The dates refer to the years in which records were made or published.

Adams, Mrs. K. F. 1959
Airy Shaw, H. K. 1941, 1945
Albertine, Sister Mary 1939
Amherst, C. 1929
Arthur, S. 1933
Babington, C. C. 1834-1851
Bailey, C. 1840-1897
Ball, J. 1836, 1849
Barker, J. W. 1847
Barthrop, N. S.
Beal, M. C.
Beavan, Miss J. 1940
Beckerlegge, J. E. 1942-1943
Bentham, G. 1858
Bennett, A. 1903
Benison, A. 1894
Bicheno, J. E. c. 1812
Bishop, E. B.
Bladon, J. 1844-1857
Bossey, F.
Bowman, J. E. c. 1830
Bree, W. T. 1824
Briggs, Miss M. A. 1957
Brody, A. O. St. c. 1860
Bruce, Archdeacon
Bryan, G. H. 1891
Buisson, Miss D. 1881
Bull, H. G. 1870
Burchell, G. 1967
Campbell, B. 1943
Charles, S. G. 1924-1951
Clarke, J. S. 1904
Clark, J. H. 1853-1868
Cobbe, Miss A. B. 1922
Cobbe, Miss M. 1922
Collett, T. G. 1956
Cooke, Miss M. C. 1933
Collins, Miss M. 1953
Conway, C. c. 1830-1837
David, Miss E. 1932
Davies, J. D. 1955
Davis, Miss K. L. 1936, 1940
Davis, Miss M. L. 1952

Dawber, Miss M. E. 1890
Dean, J. D. 1922
D'Elboux, R. H. 1923
Donovan, E. 1804
Downing, L. 1909
Druce, G. C. 1908
Duncan, 1804
Durham, H. E.
Edees, E. S. 1955-1956
Elliot, E. A. 1949
Ellis, Mrs. J. C. 1942
Evans, H. A.
Evans, T. G. 1957-1969
Foggitt, J. T.
Forster, E. 1805
Fox, H. E.
Francis, Mrs. G. C. 1912-1913
Fraser, Capt. 1918
Frederick, Miss B. M. 1928
Freer, J. 1951-1952
Gale, H. O. c. 1945
Gambier-Parry
Gee, Miss E. 1879, 1885
Gissing, T. W. 1853
Godbey, J. c. 1940
Goss, H.
Gough, J. W. 1946
Grigson, G.
Grimes, J. 1929
Guile, D. P. M. 1953
Gwilliam, B. 1940
Hall, Mrs. P. C.
Hamilton, S. c. 1900-1909
Hardaker, 1936
Harrhy, Miss I. 1940
Harrison, S. G. 1962-1968
Heeps, Miss D. 1940
Heeps, T. 1940
Holland, P. H. 1924
Hollings, Mrs. L. E. 1953
Hopkinson, M. 1956
Horne, D. W.
Hort, F. J. A. 1850

LIST OF RECORDERS

Hubbard, C. C. 1937
Hudd, A. E. 1909
Hutton, R. S. 1894
Hyde, H. A. 1932-1944
Jacques, P. 1940
James, J. W. 1927
Jeans, C. 1967
Jenkins, Miss E. A. 1931, 1936
Jenkins, Miss E. M. 1930-1936
Jones, A.
Jones, Miss F. E. 1956
Jesse, E. 1844
Jukes, H. A. L. 1927
Lamb, J. 1916
Lane, J. c. 1940
Lannon, D. 1944, 1957
Leake, Miss P. 1932
Lees, E. c. 1840-1868
Lees, F. A. 1851-1872
Leney, Mrs. J. B. 1946
Lewis, R. 1944.
Lewis, T. 1944
Ley, A. 1874-1904
Ley, Mrs. A.
Lightfoot, J. 1773
Linton, E. F. 1890
Lousley, J. E. 1933, 1935
Lovegrove, W. 1943
Lowe, E. J. 1867
Lucus, G. L. 1955
Ludlow, Mrs.
McClintock, D. 1954
Macnabb, Mrs.
Macqueen, J. 1943-1953
Manby, G. W. 1802
Marshall, E. S. 1903-1916
Mathews, W. 1849-1884
Matthews, L. H. 1932
McKenzie, A. 1922-1923
McLean, R. C. 1929
Melville, R. 1927
Millichamp, R. I. 1967
Milne-Redhead, E. 1959
Monington, H. W. 1886-1888
More, A. G. 1874

Morgan, R. H.
Motley, J. 1841
Neale, Miss H. 1953
Nelmes, E. 1945
Nelmes, W. 1941
Newman, E.
Newton, Miss S. E. 1955
Norman, Miss G. 1951
Norton, F. c. 1925
Ogden, Miss B. M. 1944
Palmer, F. 1959
Parkinson, C. c. 1894
Parris, Mrs. A. 1962
Parry, Miss C. 1937
Parry, Miss G. 1957
Parry, J. 1937
Parry, Miss M. 1937
Parry, R. 1937
Parry, T. R. Gambier
Paterson, Mrs. 1916
Perman, E. P. 1935-1937
Phillips, G. 1909
Pinkard, Mrs. A. B. 1968
Poole, J. 1837
Post, Miss E. 1927
Prout, A. 1936
Price, W. R.
Pugsley, H. W. 1948
Purchas, A. G. 1839
Purchas, W. H. 1845-1895
Purton, T. 1817-1821
Ray, J. 1662
Rees, J. 1927, 1941-1944
Reader, H. P. 1886
Redgrove, H. S.
Richards, O. W. 1920
Richards, P. W. 1920
Rickards, Miss H. 1925
Rickards, R. W. 1923-1935
Riddelsdell, H. J. 1908-1924
Rogers, F. A. 1892
Rogers, W. M. 1882
Ronnfeldt, W. 1895
Roper, Miss I. M. 1925
Rowland, A. 1938

LIST OF RECORDERS

Rowland, H. 1947-1949
Sainsbury, Miss C. 1962
Salmon, C. E. 1926
Salmon, H. E. 1925
Sandwith, N. 1946
Sankey-Barker, Mrs. C. M. 1957
Scholberg, Mrs. 1924
Sculthorpe, E. 1936
Seddon, B. 1962
Shoolbred, W. A. 1865-1920
Smith, R. L. 1924-1927
Southall, H. 1875
Storrie, J. 1886
Stratton, Miss 1957
Stumbles, R. E. 1953
Thomas, Miss E. M. 1941
Thomas, J. W. 1933
Thomas, T. H. 1850-1902
Thompson, Prof.
Thwaites, G. H. K. 1847
Townsend, C. C. 1949
Trapnell, C. G. 1921
Vachell, C. T. 1876, 1899
Vachell, Miss E. 1912, 1942
Vaughan, Mrs. I. M. 1959
Vernall, II. J. 1952-1958

Wade, A. E. 1920-1969
Walford, T. 1781
Walters, S. M. 1954
Warner, R. 1797
Watkins, B. M. 1873
Watson, W. C. R. 1933
Watt, J. 1943
Webb, J. A. c. 1945
Welch, Mrs. B. 1939
Welch, F. B. A.
Whitwell, W. 1881-1891
Willan, G. R. 1909-1911
White, J. 1886
Williams, E. H. 1940
Williams, Lewis, 1941
Williams, R. 1966
Williams, T. L. 1929
Winch, N. J. c. 1800
Wood, Dr. 1884
Woodcock, M. 1933
Woodall, Miss 1894
Woodhouse, Miss
Woods, J. 1802-1856
Worsley-Benison, F. W. S. 1881
Young, D. P. 1951

HERBARIA

J. E. Bicheno Herbarium. Formerly at the Royal Institution of South Wales. Destroyed by enemy action in 1942.

J. H. Clark Herbarium. Newport Museum and Art Gallery.

British Museum (Natural History), London.

A. Ley Herbarium. Birmingham University.

Welsh National Herbarium, National Museum of Wales. Includes the C. Conway and W. A. Shoolbred Herbaria, as well as more recent collections.

PLAN OF THE FLORA

The arrangement and the nomenclature follow the *List of British Vascular Plants* by J. E. Dandy (1958) and his Nomenclatural Changes in British Plants (1969), *Watsonia*, **7**, 157-178. Synonyms, printed in italics, are given where the accepted name may be unfamiliar to the reader. English names adopted are those given in *Welsh Flowering Plants* by H. A. Hyde and A. E. Wade, second edition (1957) and in *Welsh Ferns, Clubmosses, Quillworts and Horsetails* by H. A. Hyde, A. E. Wade and S. G. Harrison (1969). The types of habitat are those in which the species has been seen in the county, together with the degree of frequency. The botanical districts in which the species has been recorded are given, except where localities are listed. For the spelling of Welsh place-names it has been thought advisable to follow the current edition of the 1-inch Ordnance Survey Maps as these are the maps most likely to be used by the reader.

The status is given as follows:—*Native*, indigenous and not believed to owe its presence to human activity. *Denizen*, an introduced species naturalized in natural or semi-natural habitats, and not dependent upon human disturbance of the habitat for its persistence. *Colonist*, an introduced species growing only as a weed of cultivation or on disturbed ground. *Alien*, an introduced species more or less established in artificial habitats. *Casual*, an introduced species which does not persist from year to year. Localities are given for the uncommon species and are arranged as far as possible in chronological order. Approximate dates for most of the records may be found by reference to the List of Recorders. Records without an authority were made by the author. First records are given where they are of historical interest, or a date follows a record where it has some relevance to the introduction of the species into the county.

SIGNS

*   Denotes that a specimen is in the Welsh National Herbarium, National Museum of Wales.

†   Denotes that a specimen is in the J. H. Clark Herbarium, Newport Museum and Art Gallery.

!   Observed by the author, but only used for a locality after the name of another recorder.

×   A hybrid, when placed between two specific names, or before the specific epithet of a binomial.

PTERIDOPHYTA

LYCOPODIACEAE

LYCOPODIUM L.

Lycopodium selago L. Fir Clubmoss

Rocky mountain heaths; very rare. Native.

1. Tarrens in the Honddu Valley, *Ley.*

Lycopodium clavatum L. Common Clubmoss

Heaths; rare. Native.

1. Hatterral range, *Bull:* Cwm-bwchel, Llanthony, *Ley:* Priory Grove Wood, Dixton Newton, *R. Lewis.* 2. Mynydd Maen, near Risca, *T. H. Thomas:* near Abergavenny, *Hamilton.* 4. Chepstow, *Hamilton:* near Cadira Beeches, Wentwood; between Tintern and Bigsweir; above Coed Ithyl, *Shoolbred:* between St. Arvans and Devauden*, *R. Lewis.*

Lycopodium alpinum L. Alpine Clubmoss

Recorded by the Rev. A. Ley from the Honddu Valley but may refer to the Breconshire part of the valley.

EQUISETACEAE

EQUISETUM L.

Equisetum fluviatile L. Water Horsetail

   E. limosum L.

Marshes and reens; rare to locally common. Native.

1. Coalpit Hill, Fforest, near Abergavenny; by the River Monnow, above Pandy*, *Charles:* between Llwyncelyn and Pont-y-spig*, *R. Lewis:* Llanfoist. 2. Varteg, *Clark.* 3. Llangybi, *Clark:* near Maes Arthur, Castleton; Malpas. 4. Virtuous Well, Trelleck, *T. G. Evans.* 5. Near Tredegar Park; St. Brides, Wentlloog; near Llanwern.

Equisetum palustre L. Marsh Horsetail

Marshes and boggy places; frequent. Native.

1. Coalpit Hill, Fforest, near Abergavenny, *Charles:* near Pont-y-spig. 2. Pant-glas, Bedwas*. 3. Near Usk, *Clark:* near Pont-hir; Llandegfedd; near Castell-y-bwch; Plas Machen. 4. Ravensnest Wood, near Tintern; Catbrook; Trelleck Bog, *Shoolbred:* Mitchel Troy, *Charles:* St. Pierre; Trelleck, *T. G. Evans.* 5. Marshfield*.

Equisetum sylvaticum L. Wood Horsetail

Woods; very rare. Native.

1. Abergavenny district, *White:* between Grosmont Wood and Hoaldalbert*, *Charles:* Grosmont Woods*, *R. Lewis:* Llanthony, *Miss Norman.* 2. Varteg, *Clark.*

EQUISETACEAE

Equisetum arvense L.   Common Horsetail

Waste and cultivated ground, roadsides, railway tracks and banks; common. All districts. Native.

Equisetum telmateia Ehrh.   Great Horsetail
  E. maximum auct.

Wet woods, damp hedgebanks and roadsides; locally frequent. Native.
1. Abergavenny district, *White:* Honddu Valley, *Ley:* east side of Graig Syfyrddin*, *Charles*. **3.** Between St. Mellons and Llanedeyrn Mill. **4.** Piercefield Woods; Rogerstone Grange; St. Arvans; Tintern; Kilgwrrwg, *Shoolbred:* between Llanwern and Bishton. **5.** Rumney.

OSMUNDACEAE
OSMUNDA L.

Osmunda regalis L.   Royal Fern

Marshes and boggy places; very rare. Native.
1. Near Pont-y-spig; Grwyne Fawr Valley, *Ley.* **2.** Mynydd Maen, above Cwmbran*, (now extinct), *T. H. Thomas.* **4.** Near Shirenewton*, (extinct), *Shoolbred.*

DENNSTAEDTIACEAE
PTERIDIUM Scop.

Pteridium aquilinum (L.) Kuhn   Bracken

Woods, pastures, heaths, moorland and hedgebanks; very common. All districts. Native.

BLECHNACEAE
BLECHNUM L.

Blechnum spicant (L.) Roth   Hard Fern

Woods, heaths, moorland and hedgebanks on acid soils; very common. Districts 1-4. Native.

ASPLENIACEAE
PHYLLITIS Hill

Phyllitis scolopendrium (L.) Newm.   Hart's-tongue Fern

Woods, hedgebanks, walls and damp shaded stonework; frequent to locally common. Districts 1-4. Native.

ASPLENIUM L.

Asplenium adiantum-nigrum L.   Black Spleenwort

Walls, rock crevices and hedgebanks; frequent. Districts 1-4. Native.

ASPLENIACEAE

Asplenium marinum L.   Sea Spleenwort
Coastal rocks; very rare. Native.
5. Sudbrook, *T. G. Evans.*

Asplenium trichomanes L. ssp. **quadrivalens.** D. E. Meyer emend. Lovis
Maidenhair Spleenwort
Walls and rocky places; common. All districts. Native.

Asplenium viride Huds.   Green Spleenwort
Crevices of limestone and Old Red Sandstone rocks in upland districts;
very rare. Native.
1. Near Honddu Waterfall, near Tarren yr Esgob, *Bull:* on the Ffwddog;
Daren-y-cwm, Honddu Valley, *Ley:* Tarren yr Esgob*, *Ley, R. Lewis,
Charles.* **2.** Lasgarn Wood, Abersychan*, *T. H. Thomas.*

Asplenium ruta-muraria L.   Wall-rue
Walls and rocky places; common. All districts. Native.

Ceterach DC.
Ceterach officinarum DC.   Rusty-back Fern
Walls; frequent to locally common. All districts. Native.
John Ray, who visited the county in 1662, recorded this fern as 'extremely
common'.

ATHYRIACEAE
Athyrium Roth
Athyrium filix-femina (L.) Roth   Lady-fern
Moist woods; fairly common. Districts 1-4. Native.

Cystopteris Bernh.
Cystopteris fragilis (L.) Bernh.   Brittle Bladder-fern
Fissures of moist shady rocks and walls; rare to locally frequent. Native.
1. Near Abergavenny, *Purton:* Ysgyryd Fawr*, *Purchas:* Lady Park Wood;
Reddings Enclosure, *Charles:* common in the Honddu and Grwyne
Fawr Valleys. **2.** Penygarn, near Pontypool; Mynydd Maen, *T. H. Thomas:*
east face of the Blorenge, *Charles:* Craig yr Hafod, near Blaenavon,
Guile: Trefil; Cwm Mil-gatw, Dukestown. **3.** Usk, *Clark.* **4.** Wynd Cliff,
Purchas, Ley, Shoolbred: Caerwent, *Clarke:* near Tintern; Newchurch,
Shoolbred: Piercefield, *Price:* Mounton, *H. A. Evans.* Black Cliff, near
Tintern.

ASPIDIACEAE

DRYOPTERIS Adans.

Dryopteris filix-mas (L.) Schott   Common Male-fern

Woods, hedgebanks and walls; common. All districts. Native.

Dryopteris pseudomas (Wollaston) Holub. & Pouzar   Golden-scale Male-fern
D. borreri Newm.

Woods and hedgebanks; common on the Old Red Sandstone and limestone. Districts 1-4. Native.

Dryopteris carthusiana (Vill.) H. P. Fuchs   Narrow Buckler-fern
D. spinulosa (Muell.) Ktze.

Woods; rare to locally common. Native.
1. Pont-y-spig, *Ley:* Beaulieu Wood*; The Kymin; Redding's Enclosure*; Crown Woods, near Monmouth; Lady Park Wood; Grosmont Wood; Tarren yr Esgob*, *Charles.* **2.** Trevethin Wood*, *T. H. Thomas.* **3.** Newport, *Newman.* **4.** Near Trelleck Common, *Ley:* Barnets Woods*; near Chepstow*; Shirenewton; The Fedw; Wentwood*; Tintern; Bigsweir, *Shoolbred:* Piercefield Woods*, *Shoolbred,!:* Manor Wood, Whitebrook; Troy Park Wood*; Pen-y-fan, near Whitebrook*; Talycoed*, *Charles:* Trelleck Bog*, *Charles,!:* St. Arvans*, *Miss Vachell.*

Dryopteris dilatata (Hoffm.) A. Gray   Broad Buckler-fern

Woods, hedgebanks and shady mountain ledges; common. Districts 1-4. Native.

POLYSTICHUM Roth

Polystichum setiferum (Forsk.) Woynar   Soft Shield-fern

Woods and hedgebanks; fairly common in districts 1-4. Native.

Polystichum aculeatum (L.) Roth   Hard Shield-fern

Woods and hedgebanks; fairly common in districts 1, 3 and 4, unrecorded from districts 2 and 5. Native.

   var. **cambricum** (S. F. Gray)
1. Daren-y-cwm, Honddu Valley, *Ley.*

THELYPTERIDACEAE

THELYPTERIS Schmidel

Thelypteris limbosperma (All.) H. P. Fuchs   Mountain Fern
T. oreopteris (Ehrh.) Slosson

Stream sides and damp woods, especially in upland districts; common. Districts 1-4. Native.

THELYPTERIDACEAE

Thelypteris palustris Schott   Marsh Fern

Bogs and marshy thickets; very rare and probably now extinct. Native.
1. Grwyne Fawr Valley; near Pont-y-spig*, *Ley.* **4.** Shirenewton*,
Shoolbred.

Thelypteris phegopteris (L.) Slosson   Beech Fern

Moist shady places, rocky woods, streamsides and by waterfalls; rare.
Native.
1. Tarrens on both the Ffwddog and Hatterral ranges, near Llanthony;
Grwyne Fawr Valley, *Ley:* Tarren yr Esgob*, *Charles,!.* **2.** Cwm Ffrwd-
Oer, Pontypool, *E. Lees:* Machen*, *Miss Vachell.* **3.** Near Mamhilad*,
T. H. Thomas. **4.** Near Tintern*; Kilgwrrwg*, *Shoolbred:* Hael Woods,
Penallt*, *Charles.*

GYMNOCARPIUM Newm. (em. Ching)

Gymnocarpium dryopteris (L.) Newm.   Oak Fern

Thelypteris dryopteris (L.) Slosson
Rocky woods, shady rocky stream banks and by waterfalls; rare to
locally frequent. Native.
1. Honddu and Grwyne Fawr Valleys; near Pont-y-spig, *Ley:* Cwmyoy,
Bull: Honddu Waterfall, near Tarren yr Esgob, *Rep. Woolhope Club,*
1867; Tarren yr Esgob*; Grosmont Wood, *Charles.* **2.** Cwm Ffrwd-Oer,
Pontypool, *E. Lees.* **3.** Near Mamhilad*, *T. H. Thomas.* **4.** Near Tintern
Abbey, *Ray, Newman, Clark:* Llandogo Falls*, *Shoolbred:* Buckle Wood,
Tintern, *Redgrove:* Penallt; between Penallt and Whitebrook, *Charles.*

First recorded by John Ray in 1662.

Gymnocarpium robertianum (Hoffm.) Newm.   Limestone Polypody

Rocky woods and mountain screes on the Old Red Sandstone and
limestone; rare. Native.
1. Tarren yr Esgob*, *Ley, Charles,!:* Lilyrock Wood; Lady Park Wood;
Highmeadow Woods*; near Hadnock Quarry*, *Charles.* **2.** Foot of the
Blorenge, *Bowman:* near Pontypool*, *T. H. Thomas:* Craig yr Hafod,
near Blaenavon*, *Guile,!* **4.** Chepstow district, *Shoolbred:* between
Penallt and Whitebrook*, *Charles:* Rogiet; Mounton, *T. G. Evans.*

POLYPODIACEAE

POLYPODIUM L.

Polypodium vulgare L. aggr.   Common Polypody

Walls, hedgebanks, rocky places, and as an epiphyte in the forks and
along the branches of trees; common. All districts. Native.

POLYPODIACEAE

Polypodium vulgare L. sens. strict.
4. Botany Bay, Tintern*, *Hyde*.

Polypodium interjectum Shivas
P. vulgare ssp. *prionodes* Rothm.
Common in districts 3 and 4, and probably so throughout much of the county.

Polypodium australe Fée
2. Hafodyrynys*, *T. H. Thomas*. 4. Tintern Abbey*, *Walters*.

AZOLLACEAE
AZOLLA Lam.
Azolla filiculoides Lam.   Azolla
Ponds and reens; locally common. Denizen. A native of Tropical America.
4. St. Pierre Park*, *H. E. Salmon*. 5. Goldcliff*, *Dean, R. L. Smith:* between Magor and Undy, *Redgrove:* Caerleon*, *Hyde:* Llanwern, *T. G. Evans:* Peterstone Wentlloog; Whitson*; Magor*.
First recorded by J. Davy Dean in 1922.

OPHIOGLOSSACEAE
BOTRYCHIUM SW.
Botrychium lunaria (L.) Sw.   Moonwort
Pastures and heaths; rare. Native.
2. Twyn-gwyn Dingle, Pontypool*, *T. H. Thomas*. 4. Between Llangwm and Newchurch†, *Clark:* Dinham, *Lowe:* near Mounton*, *Hutton, Shoolbred:* 'The Rifle Butts,' Chepstow, *Shoolbred*.

OPHIOGLOSSUM L.
Ophioglossum vulgatum L.   Adder's Tongue
Pastures and meadows; rare to locally frequent. Native.
1. Cwmyoy, *Bull:* Staunton Road, Monmouth*; Hadnock Farm*; near Henllan, Honddu Valley*; *Charles:* Dunkard, Cross Ash, *Collett:* Lady Park Wood. 2. Pontypool Park, *Sister Albertine*. 3. Usk†; Prioress Mill†; Rhadyr†, *Clark*. 4. Near Tintern, *Gissing:* Llangeview†, *Clark:* St. Arvans*, *Hutton:* Barnets Woods*; near the Barnets Farm*; Piercefield Park; between Mounton and Dinham; Shirenewton, *Shoolbred:* Little Llanthomas, *Miss C. Parry:* The Minnetts and Slade Wood*; Rogiet, *Mrs. Ellis:* Ringland Top, near Llanwern; by Trelleck Bog*.

SPERMATOPHYTA
GYMNOSPERMAE
PINACEAE
PINUS L.

Pinus sylvestris L. Scots Pine

Denizen.

4. A few self-sown trees occur on Trelleck Bog*.

TAXACEAE
TAXUS L.

Taxus baccata L. Yew

Abundant in woods on the limestone in the Wye Valley, elsewhere as a planted tree. Native.

ANGIOSPERMAE
DICOTYLEDONES
RANUNCULACEAE
CALTHA L.

Caltha palustris L. Marsh Marigold

Marshes, wet meadows and borders of ponds and streams; common all districts. Native.

TROLLIUS L.

Trollius europaeus L. Globe Flower

River banks and marshy meadows in upland districts; rare. Native.
1. Grwyne Fawr Valley*, *Ley, Mrs. Paterson.* **2.** Near Pontnewydd; Varteg*, *Conway†, Clark:* Risca*, *C. T. Vachell.* **3.** Near Michaelstone y Vedw*, *Smith.*

HELLEBORUS L.

Helleborus foetidus L. Fetid Hellebore

Woods and thickets on calcareous soils; rare. Native.
1. Near Monmouth, *Hamilton, Charles.* **2.** Cwm Ffrwd, Abersychan, *Freer.* **4.** The Coombe, near Llanfair Discoed*, *Shoolbred, Hamilton, Miss Rickards:!.* Wynd Cliff Woods, *Shoolbred.*

51

RANUNCULACEAE

Helleborus viridis L. ssp. **occidentalis** (Reut.) Schiffn.   Green Hellebore
Woods and copses; rare. Native.
1. By the River Wye, near Monmouth, *Lees, Hamilton, Charles*. **3.**
Prescoed†, *Clark:* near Chain Bridge†; near Kemeys Folly*†, *Clark,
Miss Rickards:* Llanbadoc, *J. Rees.* **4.** Magor; wood near Llanfihangel,
Dr. Clarke: St. Bride's Netherwent, *Mrs. Ellis:* Leasbrook*, *Miss Rickards.*

ACONITUM L.

Aconitum anglicum Stapf   Monkshood
Locally common by streams and rivers. Native.
1. Llangattock Lingoed, *Bladon:* Llangattock Vibon Avel, *Charles.*
3. Pont-rhyd-yr-ynn, *Conway:* Llanbadoc†; Cefn-ila, *Clark:* near
Michaelstone y Vedw, *Richards:* near Pentwyn Farm, Llanddewi Fach,
Campbell: Llantarnam Abbey, *Rees:* by the River Rhymney, St. Mellons*.
4. Mounton Valley, *Hamilton, Shoolbred:* Itton*†, *Clark, Shoolbred:*
Shirenewton*, *Ley, Shoolbred:* near Rhyd-y-fedw, *Shoolbred:* between
Shirenewton and Grey Hill, *H. A. Evans:* Llwynycelyn, *T. G. Evans.*

DELPHINIUM L.

Delphinium ambiguum L.   Larkspur
D. gayanum Wilmott; *D. ajacis* auct.
Casual.
4. Caerwent*, *Hudd.* **5.** In an orchard, Marshfield, *Woodcock.*

ANEMONE L.

Anemone nemorosa L.   Wood Anemone
Common in woods, hedgerows and pastures. All districts. Native.
Pink-flowered forms are not uncommon.

Anemone apennina L.   Blue Mountain Anemone
Garden escape.
3. Hedgerow, St. Mellons*, *Lovegrove,* 1943.

CLEMATIS L.

Clematis vitalba L.   Traveller's Joy
Hedgerows, thickets and woods. Common on calcareous soils, frequent
on the Old Red Sandstone. Districts 1-4. Native.

RANUNCULUS L.

Ranunculus acris L.   Meadow Buttercup
Common in meadows, pastures, on grassy roadsides and woodland rides.
All districts. Native.

RANUNCULACEAE

Ranunculus repens L.   Creeping Buttercup

Abundant in cultivated ground and on roadsides, common in meadows and hedgerows. All districts. Native.

Ranunculus bulbosus L.   Bulbous Buttercup

Common in pastures, meadows and on roadsides. All districts. Native.

Ranunculus arvensis L.   Corn Crowfoot

Cultivated ground; rare. Colonist.
1. Abergavenny, *Hamilton:* Fiddler's Elbow, Monmouth*; Onen*; Llangattock Vibon Avel*, *Charles:* Watery Lane, Monmouth*. **3.** Castleton; Michaelstone y Vedw; Llantarnam, *Hamilton:* Usk district, *Miss Frederick:* near the Wilderness, Malpas. **4.** Near Chepstow, *Shoolbred.*

Ranunculus sardous Crantz   Hairy Crowfoot

Moist pastures, usually near the coast; rare. Native.
1. Abergavenny, *Hamilton.* **3.** Pontypool Road, *Hamilton.* **5.** Dyffryn, Nash and Llanwern, *Hamilton.*
The records for districts 1 and 3 may be errors.

Ranunculus parviflorus L.   Small-flowered Crowfoot

Dry banks; very rare. Native.
1. Near Monmouth, *Hamilton.* **3.** Near Pontnewydd Works, *Conway:* railway embankment near Usk, *Clark.*

Ranunculus auricomus L.   Goldilocks

Locally frequent in hedgerows and woods. Native.
1. Lilyrock Wood and Garth Wood, near Monmouth*; near Henllan, Honddu Valley*; Graig Syfyrddin*; Dingestow*; Tregare*; Halfway House Wood*; Troy Farm; near St. Maughan's*; Lady Park Wood; between Wyesham and The Kymin; near May Hill Wharf*, *Charles:* Seargent's Gorse, Monmouth; between Abergavenny and Llantilio Pertholey; between Pandy and the Toll House*; Llanvihangel Crucorney. **2.** Pontypool*, *T. H. Thomas.* **3.** Henllys Wood and near The Coldra, *Hamilton:* Usk district, *Miss Frederick:* between Castleton and Michaelstone y Vedw. **4.** Wynd Cliff*; *Clark, Shoolbred:* Dinham Woods; Runstone; Mounton, *Shoolbred:* Tal-y-coed; Upper Redbrook*, *Charles:* Rogiet, *Mrs. Ellis:* near Carrow Hill.

Ranunculus lingua L.   Greater Spearwort

Marshes; rare. Native.
2. Machen, *Riddelsdell.* **4.** Near Trelleck, *T. G. Evans.*

RANUNCULACEAE

Ranunculus flammula L.   Lesser Spearwort
Common in marshes and by streams. All district. Native.

Ranunculus sceleratus L.   Celery-leaved Crowfoot
Locally common in marshes and on margins of ponds, streams and reens. Native.
1. Wonastow, *Charles*. **3.** Michaelstone y Vedw; Caerleon; Malpas, *Hamilton:* Usk, *Clark, Miss Frederick:* Castleton. **4.** Plentiful on the banks of the River Wye and ponds in the Chepstow district; Mathern*; Sudbrook, *Shoolbred*. **5.** Common throughout.

Ranunculus hederaceus L.   Ivy-leaved Water Crowfoot
Occasional on the muddy margins of ponds. Native.
2. Risca*, *C. T. Vachell:* Trefil; near Blaen-y-cwm, Dukestown. **3.** Near Foxwood, Malpas, *Hamilton:* Castleton*; Henllys. **4.** Chepstow, *Shoolbred:* Trelleck, *Charles:* Catbrook, *T. G. Evans.* **5.** Marshfield, *Hamilton:* Rumney*.

Ranunculus omiophyllus Ten.   Lenormand's Water Crowfoot
 Ranunculus lenormandi F. W. Schultz
Common in the upland districts 1 and 2, rare elsewhere. Native.
3. By Coed Mawr, near Rhiwderyn*; between Castleton and Michaelstone y Vedw; near Court Perrott, Llandegfedd. **4.** Trelleck Bog, *Watkins, Charles.*

Ranunculus fluitans Lam.   River Water Crowfoot
Rivers and fast flowing streams; locally common. Native.
1. River Wye from Monmouth down to within a mile of the tidal part*, *Shoolbred, Charles:* Hadnock stream, near Monmouth*, *Charles.*

Ranunculus circinatus Sibth.   Fan-leaved Water Crowfoot
Reens; very rare. Native.
5. Reen near Magor Pill Farm*, *Airy-Shaw,!:* Blackwall Reen, Magor.

Ranunculus trichophyllus Chaix ex Vill.   Fennel-leaved Water Crowfoot
Reens, ponds and slow-flowing streams; rare. Native.
4. Caerwent, *Shoolbred.* **5.** Near Magor, *Shoolbred,!:* Mathern, *T. G. Evans:* Rumney*; Peterstone Wentlloog*; Undy.

Ranunculus aquatilis L.   Water Crowfoot
 Ranunculus heterophyllus Weber
Ponds and ditches; frequent. Native.
1. Llanfoist*; near Troy Station, Monmouth*. **4.** Broadwell; St. Pierre;

RANUNCULACEAE

near Trelleck; Crossway Green Farm, Chepstow*; near the Wynd Cliff*; Barnet's Farm, Chepstow*, *Shoolbred:* Wentwood Reservoir. **5.** Marshfield*, *Miss Vachell:* Undy*; Magor*, *Shoolbred:* near Peterstone Wentlloog*.

Ranunculus peltatus Schrank   Water Crowfoot

Ponds and reservoirs; locally frequent. Native.
1. Llanfoist*. **2.** Pontypool, *Bladon.* **3.** Coed-y-paen*, *Conway.* **4.** Chepstow district; Portskewett*; The Glynn, Itton*; Llanishen, *Shoolbred:* Wentwood Reservoir, *Mrs. Parris:* Fairoak Pond*, *T. G. Evans,!:* Penallt Common.

Ranunculus penicillatus (Dumort.) Bab.
  Ranunculus pseudofluitans (Syme) Newbould
Rare. Native.
3. Pontnewydd Pond*, *Conway.*

Ranunculus baudotii Godr.   Baudot's Water Crowfoot

Frequent in ponds and ditches near the sea. Native.
4. Ponds by the Wye, Chepstow; The Meads, Chepstow*, *Shoolbred.*
5. Undy Marshes*; moors below Mathern*, *Shoolbred:* Peterstone Wentlloog, *Miss Vachell:* Caldicot Level, *T. G. Evans:* Rumney*; Newport Docks*.

Ranunculus ficaria L.   Lesser Celandine

Very common in meadows, grassy roadsides, hedgerows, woods and cultivated ground. All districts. Native.
ssp. **bulbifer** (Marsden-Jones) Lawalrée
4. Chepstow*, *Shoolbred.*

ADONIS L.

Adonis annua L.
  A. autumnalis L.   Pheasant's Eye
Cultivated ground; a rare casual.
4. Magor, *Clark:* near Chepstow*, *Shoolbred.*

AQUILEGIA L.

Aquilegia vulgaris L.   Columbine

Woods, thickets, hedgerows and railway embankments; common in district 4, frequent in districts 1 and 3. Native.
1. Abergavenny, *White:* about Monmouth*, *Hamilton, Charles, Lewis.*
3. near Monkswood Chapel; by Berthin Brook; near Coed Cwnwr, *Clark:* near Castell-y-bwch; near Coed Garw, Bettws; near Coed Mawr, Goytre; Pant-yr-eos, Gwehelog*; between St. Mellons and Michaelstone y Vedw; near Ynys-y-fro*.

RANUNCULACEAE

THALICTRUM L.

Thalictrum flavum L.   Meadow Rue

Margins of streams, reens and rivers; rare. Native.
1. Banks of the Wye, between Monmouth and Dixton*; Hadnock*; by the River Monnow, Rockfield*, *Charles.* **3.** Olway Brook, near Usk†, *Clark.* **4.** About Tintern*; Bigsweir; Llandogo, *Shoolbred:* near Llandevaud; River Wye, Penallt*; Redbrook*; *Charles:* near Rogiet*, *Mrs. Ellis:* River Wye, above Chepstow, *T. G. Evans:* Llanbeder*. **5.** Reens near Magor, *Hamilton, Shoolbred.*

BERBERIDACEAE

BERBERIS L.

Berberis vulgaris L.   Barberry

Hedgerows and borders of woods; rare. Denizen.
1. Wonastow Road, near Monmouth*; Hadnock Farm, Monmouth*, *Charles.* **3.** Wood at Trostrey Weir, *Clark:* Bulmoor, Caerleon, *Hamilton:* Llanddewi Fach*. **4.** Near Tintern, *Gissing:* wood near Earlswood Common, *Clark:* Wye Valley, near Monmouth, *Hamilton:* Chepstow Castle*, *Shoolbred.*

MAHONIA Nutt.

Mahonia aquifolium (Pursh.) Nutt.   Oregon Grape
   Berberis aquifolium Pursh.

Plantations and borders of woods; planted and naturalized.
1. Plantation above Garth, Staunton Road, *Charles:* among brambles on the edge of Garth Wood*, *R. Lewis.*

NYMPHAEACEAE

NYMPHAEA L.

Nymphaea alba L.   White Waterlily

Rivers and ponds; rare. Native.
1. Abergavenny, *White:* River Trothy, near Monmouth; lake at The Hendre (? introduced), *Hamilton.*

NUPHAR Sm.

Nuphar lutea (L.) Sm.   Yellow Waterlily

Ponds and canals; rare. Native.
1. Llanfoist. **2.** Canal between Cwmbran and Pontnewydd, *A. Jones.* **3.** Coldra, near Christchurch; Llanwern, *Hamilton.*

CERATOPHYLLACEAE

CERATOPHYLLUM L.

Ceratophyllum demersum L.   Hornwort

Reens and canals; rare. Native.
1. Canal near Abergavenny, *Bailey*. 5. Near Peterstone Wentlloog*, *Harrison & Seddon:* Marshfield*; near Magor Pill*.

Ceratophyllum submersum L.

Reens; rare. Native.
5. Whitson, *T. G. Evans:* Marshfield*; Near Peterstone Wentlloog*.

PAPAVERACEAE

PAPAVER L.

Papaver rhoeas L.   Red Poppy

Cutlivated ground and disturbed soil on roadsides; frequent to common in districts 1 and 4, occasional elsewhere. Native.

Papaver dubium L.   Smooth Long-headed Poppy

Cultivated and waste ground; frequent in districts 1, 3 and 4, rare elsewhere. Native.
1. Cornfields about Monmouth*; Tregare*, *Charles:* Pont-y-spig*.
2. Near Maes-y-cwmmer*. 3. Llancayo; Monkswood; Estavarney†; near Chain Bridge†, *Clark:* Raglan district, *Miss Frederick:* near Michaelstone y Vedw; Fairwater*; near Penylan, Llandegfedd*. 4. Trelleck†*, *Clark*, *Charles:* Wye Valley, *Hamilton:* Chepstow district; Portskewett*, *Shool-bred:* near the Severn Tunnel, *Druce:* Upper Redbrook*, *Charles:* Tymawr, Penallt, *Beckerlegge*. 5. Between Castleton and Marshfield*.

Papaver argemone L.   Rough Long-headed Poppy

Cultivated ground; very rare. Colonist.
4. Between St. Arvans and Chepstow Road*, *Shoolbred*.

Papaver somniferum L.   Opium Poppy

Waste ground; rare. Casual.
1. Monmouth, *Charles:* Cleddon*. 2. Croesyceiliog, near Pontnewydd*.
4. Chepstow district, *Shoolbred:* Trelleck; Whitebrook*, *Charles*.

MECONOPSIS Vig.

Meconopsis cambrica (L.) Vig.   Welsh Poppy

Mountain rocks and rocky banks of rivers; rare and restricted in range. Native.

PAPAVERACEAE

1. On the tarrens of the Honddu Valley just on the border of Breconshire and Monmouthshire and extending into both counties; lower part of the Grwyne Fawr Valley, *Ley*. A specimen collected by Dr. C. T. Vachell and labelled 'near Abergavenny' doubtless came from the latter locality.

GLAUCIUM Mill.

Glaucium flavum Crantz   Horned Poppy

Shingle by the River Severn; rare. Native.
4. Chepstow, *Hamilton*. 5. Black Rock, *Bicheno:* near Mathern Pill*, *Shoolbred.* By the River Usk, Newport†, *Clark.*

CHELIDONIUM L.

Chelidonium majus L.   Greater Celandine

Roadsides, hedgerows and old walls, usually near habitations; frequent and widely distributed. All districts. Denizen.

FUMARIACEAE

CORYDALIS Medic.

Corydalis claviculata (L.) DC.   Climbing Fumitory

Rocky woods, thickets and gravelly hillsides; locally frequent. Native.
1. Abergavenny district, *White's Guide*. 2. Near Abercarn, *McKenzie:* Cwm Gwyddon; Cefn Rhyswg, near Abercarn*; near Wattsville*. 3. Craig y Wenallt, Henllys*. 4. Near Chepstow, *St. Brody:* Wynd Cliff, *Shoolbred:* Trelleck*; *Perman, Charles:* Botany Bay, Tintern*, *Perman*, near Cae Jack, Penallt*, *Charles:* Tymawr, Penallt*, *Beckerlegge:* Wentwood, *Sandwith:* Llandogo.

Corydalis lutea (L.) DC. Yellow Fumitory

Old walls; garden escape. Frequent.
1. Llangattock-Vibon-Avel*; Kymin Hill, Monmouth*; Monmouth Cap*; near Llanvihangel Crucorney*; near Norton*; Onen*; Buckholt*; Bailey Pit Farm, near Monmouth*; Cwmyoy*; Grosmont*, *Charles:* Llanfoist. 3. Kemeys Commander†; Usk, *Clark:* Pencarn, *Hamilton:* Usk district, *Mrs. Hall*. 4. Chepstow†, *Clark, Shoolbred:* The Glyn*, *Shoolbred:* New Mills, near Penallt*; Upper Redbrook*, *Charles:* Whitelye; Llandogo; about Tintern. 5. Dyffryn, *Hamilton.*

FUMARIA L.

Fumaria purpurea Pugsl.   Purple Fumitory

Cultivated ground and river banks; rare. Colonist.
1. Kymin Hill, Monmouth*, *Charles*. 3. Banks of the River Usk, Usk*, *Mrs. Scholberg*. 4. Tintern, *Miss Todd.*

Fumaria bastardii Bor.   Tall Fumitory

Cutlivated ground; rare. Colonist.

4. Llandogo, *Ley.*

Fumaria muralis Sond. ex Koch ssp. **boraei** (Jord.) Pugsl. Boreau's Fumitory

Cultivated ground; rare. Colonist.

3. Llanrumney*. **4.** Llandogo*, *Shoolbred.* **5.** Caldicot*, *Mrs. Ellis:* between Castleton and Marshfield*; St. Mellons*.

Fumaria officinalis L.   Common Fumitory

Cultivated ground; frequent. Colonist.

1. Kymin Hill, Monmouth*; Hadnock Farm, near Monmouth, *Charles:* near Abergavenny*; Pont-y-spig. **3.** Newport, *Hamilton:* Raglan district, *Miss Frederick:* Usk*, *Charles, Miss Vachell:* Llanrumney*; Bassaleg; Llandegfedd*. **4.** Severn Tunnel, *Hamilton:* Chepstow; Ifton*; Rogiet*; Caerwent*, *Shoolbred.*

CRUCIFERAE
BRASSICA L.

Brassica oleracea L.   Cabbage

Rocks, walls and river banks; rare. A doubtful native or relic of cultivation.

3. Banks of the River Usk, Newport, *Clark.* **4.** Rocks below Chepstow Castle, *Lightfoot, Banks, Clark, Shoolbred, Trapnell,!.* **5.** Llanwern, garden escape.

First recorded by John Lightfoot in 1773.

Brassica napus L.   Rape

Waste ground, roadsides and river banks. Frequent as a relic of, or escape from, cultivation. All districts.

Brassica rapa L. var. **campestris** (L.) Koch   Wild Turnip

River banks and roadsides; locally frequent. Colonist or denizen.

1. Banks of the River Wye, Monmouth; banks of the River Monnow, Skenfrith, *Charles:* Honddu and Grwyne Fawr Valleys, *Ley.* **3.** Castleton, *Hamilton:* Caerleon*. **4.** Banks of the River Wye above Tintern, Bigsweir* and Llandogo; St. Kinsmark*, *Shoolbred.*

Brassica nigra (L.) Koch   Black Mustard

Banks of rivers and reens, roadsides, waste ground and railway tracks; locally common. Native.

CRUCIFERAE

1. Abergavenny; Llanthony, *Hamilton:* Monmouth*, *Hamilton, Charles:* Skenfrith, *Charles:* Llanvapley, *J. Rees.* **4.** Banks of the River Wye, Chepstow, *Watkins, Hamilton, Shoolbred:* Brockweir, *Shoolbred:* Bigsweir*, *Watkins, Shoolbred:* Whitebrook*, *Charles.* **5.** Near Newport, *Hamilton:* near Rogiet*, *Shoolbred:* Traston; Newport Docks; Rumney*.

RHYNCHOSINAPIS Hayek

Rhynchosinapis cheiranthos (Vill.) Dandy   Channel Isles Cabbage
Rare. Alien.
2. Disused railway track below Mynydd Dimlaith, 1968*.

SINAPIS L.

Sinapsis arvensis L.   Charlock
Brassica arvensis (L.) Rabenh.

A common weed of cultivated ground, roadsides and waste ground. All districts. Colonist.

Sinapis alba. L.   White Mustard
Brassica alba (L.) Rabenh.

Waste ground; rare. Casual.
1. Monmouth*; Chippenham*, *Charles.* **4.** Wye Valley, *Hamilton:* above Tintern, *Ley.*

HIRSCHFELDIA Moench

Hirschfeldia incana (L.) Lagr. Foss.   Hoary Mustard
Brassica incana (L.) F. Schultz; Brassica adpressa Boiss.

Waste ground and roadsides; rare. Alien.
5. Newport Docks*, *Macqueen:* Goldcliff, *J. N. Davies:* coast road, Rumney*.

DIPLOTAXIS DC.

Diplotaxis muralis (L.) DC.   Sand Rocket

Roadsides and railway tracks; locally frequent. Established alien.
1. Near Mayhill Station, Monmouth*; Hadnock*, *Charles.* **4.** Railway track from Chepstow to Magor*; Tintern*; Portskewett, *Shoolbred:* near Redbrook Railway Station*, *Charles.* **5.** Severn Tunnel Junction*; Llanwern*.
Less frequent than formerly owing to the use of weed killers on railway tracks.

Diplotaxis tenuifolia (L.) DC.   Wall Rocket

Waste ground, railway tracks and river banks; locally frequent. Established alien.

CRUCIFERAE

3. Banks of the River Usk, Newport * †, *Conway*, 1834, *Clark, Hamilton, Whitwell:* railway bank, Newport, *Hamilton*. **5.** Newport Docks, *Hamilton,!:* Spencer Steel Works*.

RAPHANUS L.

Raphanus raphanistrum L.   Wild Radish

Cultivated ground; common. Districts 1, 3 and 4. Colonist.

CONRINGIA Adans.

Conringia orientalis (L.) Dumort.
Erysimum orientale (L.) R.Br.

Waste ground; rare. Casual.
1. Kymin Hill, Monmouth*, *Charles*. **4.** Portskewett*, *Shoolbred*.

LEPIDIUM L.

Lepidium campestre (L.) R.Br.   Field Pepperwort

Cultivated ground, hedgerows, quarries, waste places, and dry pastures; rare. Native.
1. Abergavenny; Llanthony, *Hamilton:* Hadnock, *Charles*. **3.** Pontnewydd*, *Conway:* near Raglan*, *Bryan*. **4.** Wye Valley, *Hamilton:* Shorncliff Woods; Mounton; Dinham; near Tintern*, *Shoolbred*. **5.** Liswerry; Newport Docks, *Hamilton*.

Lepidium heterophyllum Benth.
Lepidium smithii Hook.

Hedgerows, roadsides and railway tracks; rare. Native.
1. Abergavenny, *Hamilton, Clark:* Wyesham*, *Charles*. **3.** Mescoed Mawr, Bettws*. **4.** Hardwick, *Clark:* Between Devauden and Kilgwrrwg*; near Fairoak, Tintern*, *Shoolbred:* Bulwark, Chepstow, *Miss M. Cobbe:* Ceciliford*, *R. Lewis*.

Lepidium ruderale L.   Narrow-leaved Pepperwort

Waste ground and roadsides; rare, Casual.
4. Chepstow, *F. A. Lees, Clark, Shoolbred:* Devauden*, *Shoolbred*. **5.** Newport Docks, *Hamilton, Smith:* Rumney*.

Lepidium neglectum Thell.

Waste ground and roadsides; rare. Casual.
1. Dixton*; Kymin Hill, Monmouth*, *Charles*. **4.** Redbrook*, *Charles*.

Lepidium latifolium L.   Dittander

Waste ground; rare. Alien.
5. Alexandra Dock, Newport, 1942*, *Macqueen*.

61

CRUCIFERAE

Lepidium densiflorum

Waste ground; rare. Casual.
4. Near the River Wye, Chepstow, *C. E. Salmon.*

Lepidium virginicum L.

Waste ground; rare. Casual.
4. Bulwark, Chepstow, *Miss A. B. Cobbe.* **5.** Alexandra Dock, Newport*,
Macqueen.

CORONOPUS Zinn

Coronopus squamatus (Forsk.) Aschers.   Swine's Cress
Senebiera coronopus (L.) Poir.

Muddy river banks, farm tracks, cultivated ground and waste places;
frequent. Native.
1. Monmouth*; cornfield, Hadnock Farm; Rockfield*; Little Chippenham*, *Charles.* **2.** Near Maes-y-cwmmer. **3.** Newport*, *Conway, Clark,
Whitwell:* St. Woolas; Bassaleg, *Hamilton:* Raglan district, *Miss Frederick:*
Castleton; Llandegfedd*. **4.** Chepstow, *Clark, Shoolbred:* near Pen-y-
clawdd, *Miss Davis:* Whitebrook*, *Charles.* **5.** Alexandra Dock, Newport,
Hamilton: near Magor Pill*; Rogiet,* *Shoolbred:* near Peterstone
Wentlloog*; Rumney.

Coronopus didymus (L.) Sm.   Wart Cress

Waste ground and roadsides; frequent and widespread. Districts 1, 3-5.
Established alien first recorded in 1868 by Clark from Newport.

CARDARIA Desv.

Cardaria draba (L.) Desv.   Hoary Cress
Lepidium draba L.

Roadsides, railway tracks, cultivated ground and waste places; locally
common. Established alien. First recorded in 1837 by Conway.
1. Monmouth*, *Charles.* **3.** Usk, *Rickards:* New Inn, Pontypool, *Miss
Jones:* Penpergwm; Malpas Road, Newport*. **4.** Chepstow*, *Conway,
Ley, Shoolbred:* Severn Tunnel Junction, *Mrs. Ellis.*

IBERIS L.

Iberis umbellata L.   Candytuft

Waste ground and railway ballast; rare. Casual.
3. Croesyceiliog, Pontnewydd*. **4.** Tintern*, *Shoolbred.*

CRUCIFERAE

THLASPI L.

Thlaspi arvense L.   Penny-Cress

Cultivated ground, roadsides and waste ground; locally frequent. Colonist.
1. Abergavenny*, *Hamilton,!:* Hadnock*; Duffield's Farm, Monmouth*;
May Hill, Monmouth*, *Charles.* **3.** Raglan district, *Miss Frederick:*
between Usk and Gwernesey, *Charles:* near Pen-y-lan, Llandegfedd,
Campbell: Castleton*. **4.** Crossway Green, Chepstow, *Clark:* Wye
Valley, *Hamilton:* Mounton; Rogerstone Grange*; Chepstow Park;
Caerwent; St. Arvans, *Shoolbred.* **5.** Alexandra Dock, Newport, *Hamilton.*

CAPSELLA Medic.

Capsella bursa-pastoris (L.) Medic.   Shepherd's Purse

An abundant weed of cultivated ground, waste ground and waysides.
All districts. Native.

COCHLEARIA L.

Cochlearia officinalis L.   Scurvy-grass

Banks of tidal rivers; rare. Native.
4. Chepstow, *Lees, Shoolbred.* **5.** Denny Island, *Mrs. M. L. Davis.*
Hamilton's record from the mouth of the river Usk, near Pilot's Pill,
may have been an error for *Cochlearia anglica.*

Cochlearia danica L.   Danish Scurvy-grass

A plant of rocky places near the sea, but the only certain record for the
county is from ballast.
5. Ballast heap, Newport*, *Conway:* River Usk Newport, *Clark,* a
probable error since the specimen so named in his herbarium is *Cochlearia
anglica.*

Cochlearia anglica L.   English Scurvy-grass

Muddy river estuaries; locally common. Native.
3. Caerleon†, *Clark:* near Malpas*. **4.** Chepstow*†, *Clark, Shoolbred:*
abundant on the banks of the River Wye for two or three miles upwards
from the mouth of the river, and the banks of the Severn, *Shoolbred:* near
Tintern, *Ley.* **5.** Newport*†, *Conway, Clark:* Nash; St. Bride's Wentlloog,
Hamilton: Marshfield, *Richards:* Rumney*, *Richards,!.*
var. **stenocarpa** Meyer
4. Banks of the River Wye, below the Wynd Cliff*, *W. H. Purchas.*

63

CRUCIFERAE

LOBULARIA Desv.

Lobularia maritima (L.) Desv.   Sweet Alyssum
Alyssum maritimum (L.) Lam.

Waste ground; rare. Casual.
5. Newport Docks, *Hamilton.*

BERTEROA DC.

Berteroa incana (L.) DC.
Alyssum incanum L.

Cultivated ground; rare. Casual.
2. Garden weed, New Tredegar, *J. W. Thomas.*

EROPHILA DC.

Erophila verna (L.) Chevall.   Whitlow-grass
Walls, limestone rocks and dry banks on basic soils; locally common.
Native.
1. Abergavenny, *Hamilton:* Monmouth, *Charles.* **2.** Trefil*. **3.** Trostrey†;
Llandenny†; Pantycollin†, *Clark.* **4.** Trelleck†, *Clark:* Chepstow; Cross-
way Green*; Mathern*; Portskewett*, *Shoolbred:* Penallt, *Charles,!:*
Rogiet, *T. G. Evans.* **5.** Marshfield, *Richards.*

Erophila spathulata Láng
Erophila boerhaavii (Van Hall) Dumort.

Limestone walls and quarries; rare. Native.
1. Hadnock*, *Charles.* **4.** Portskewett, *Marshall:* Tintern*, *Druce, H. E.
Fox:* near Tintern, *Shoolbred:* Cleddon*; Trelleck*, *Charles.*

ARMORACIA Gilib.

Armoracia rusticana Gaertn., Mey & Scherb.   Horse Radish
Cochlearia armoracia L.

Roadsides, railway banks and waste ground; frequent and widespread.
All districts. Alien.

CARDAMINE L.

Cardamine pratensis L.   Lady's-smock
Pastures, meadows, woods, banks of streams and marshes; common and
widespread. All districts. Native.

Cardamine amara L.   Large Bitter-Cress
Marshes; very rare. Native.
4. Between Mounton and Penterry, *Shoolbred.*

CRUCIFERAE

Cardamine impatiens L.   Impatient Bitter-Cress

Rocky banks, quarries and woodland paths on calcareous soils; locally frequent. Native.
1. Wyesham, near Highmeadow siding*; Lady Park Wood; Hadnock Wood*; near May Hill Station, Monmouth*; Beaulieu Wood*, *Charles.* **2.** Abergavenny, *Hamilton.* **4.** About Tintern*, *Forster, Gissing, Shoolbred:* Hardwick Clifft; Chepstowt, *Clark:* Bigsweir; Llandogo, *Shoolbred:* Piercefield, *Price:* Wynd Cliff, *Redgrove,!:* Penallt*, *Charles:* Black Cliff, near Tintern*.

Cardamine flexuosa With.   Wood Bitter-Cress

Woods, about wet rocks and wall bases, and banks of ditches; common. All districts. Native.
×**pratensis**
2. near Pontnewydd*.

Cardamine hirsuta L.   Hairy Bitter-Cress

Walls, rocks, path sides and banks; frequent to common. All districts. Native.

BARBAREA R.Br.

Barbarea vulgaris R.Br.   Common Winter-Cress

Frequent and widespread on stream and river banks, waste places and roadsides. Districts 1, 3 and 4. Native.

Barbarea intermedia Bor.   Intermediate Winter-Cress

Cultivated and waste ground; rare. Colonist.
1. Near Cwmyoy, *Ley.* **2.** Near Llantarnam*; near Walnut Tree Farm, Llandegfedd*.

Barbarea verna (Mill.) Asch.   American Winter-Cress

Cultivated ground, river banks and railway embankments; rare. Colonist.
1. Bank of the River Wye, Monmouth*; Kymin Hill, Monmouth*, *Charles.* **3.** Pontypool Road*. **4.** St. Arvans*; Chepstow, *Shoolbred:* near Pen-y-clawdd, *Parry:* near Trelleck*.

ARABIS L.

Arabis hirsuta (L.) Scop.   Hairy Rock-Cress

Limestone rocks, banks and walls; locally common. Native.
1. Lady Park Wood, *Charles,!.* **4.** Tinternt, *Forster, Clark, Shoolbred:* Wynd Cliff*, *E. Lees, Clark, Hamilton, Shoolbred, Charles,!.* Hardwick Clifft, *Clark:* Chepstow Castle and cliffs; between Trelleck and Tintern, *Shoolbred:* between Highmoor Hill and Caerwent, *Mrs. Ellis.*

65

CRUCIFERAE

TURRITIS L.
Turritis glabra (L.) Bernh.
Waste ground. Casual.
5. Alexandra Dock, Newport, *Hamilton*.

NASTURTIUM R.Br.
Nasturtium officinale R.Br. *sens. lat.* Watercress
Streams, springs and other running water; common. All districts. Native.
The following segregates are recorded.
N. officinale R.Br. *sens. strict.*
1. Pont-y-spig*. **4.** Near Tintern, *Townsend.* **5.** Rumney*.

N. microphyllum (Boenn.) Reichb.
1. Near Pont-y-spig; Abergavenny; Dingestow. **3.** Plas Machen. **4.** Mounton*, *Shoolbred:* Fairoak Pond. **5.** Rumney*; Peterstone Wentlloog*; Undy*.

RORIPPA Scop.
Rorippa sylvestris (L.) Besser Creeping Yellow-Cress
Nasturtium sylvestre (L.) R.Br.
River banks, reen-sides, waste ground and cultivated ground; widespread but not common, unrecorded from District 2. Native.
1. Monmouth*, *Charles.* **3.** Pencarn, *Hamilton:* Usk*, *Charles:* Raglan district, *Miss Frederick.* **4.** Bigsweir, *Ley:* Tintern*; Brockweir*; Chepstow*, *Shoolbred:* Llandogo*, *Shoolbred, Mrs. Leney:* Whitebrook*; Redbrook*, *Charles.* **5.** Marshfield, *Hamilton:* Newport, *Clark, Hamilton:* Rumney*.

Rorippa palustris (L.) Besser Yellow-Cress
R. islandica auct. *Nasturtium palustre* (L.) DC.
Margins of ponds, river banks and reen-sides; frequent. Native.
1. Abergavenny*; Monmouth; Llanfoist*; Wonastow*, *Charles:* Dingestow. **3.** About Usk*†, *Clark, Charles:* Llantarnam, *Hamilton:* by the Afon Llwyd, Llangattock. **4.** Between Tintern Cross and the Fedw*; Fairoak; near Tintern; Newton Green; Mathern*; St. Pierre Pond, *Shoolbred:* Caerwent Brook, *Mrs. Ellis:* Ifton. **5.** Coedkernew, *Hamilton:* Magor, *Lewis:* Peterstone Wentlloog; Newport Docks*.

Rorippa amphibia (L.) Besser Water Rocket
Nasturtium amphibium (L.) R.Br.
Marshes, river banks and canals; rare. Native.
1. Monmouth*; Dixton*; Hadnock*, *Charles.* **2.** Between Newbridge and Crumlin, *McKenzie.* **3.** Malpas*. **4.** Banks of the Wye above Tintern*, *Shoolbred:* Bigsweir; near Brockweir*; Penallt*, *Charles.* **5.** Undy*.

CRUCIFERAE

HESPERIS L.

Hesperis matronalis L.   Dame's Violet

Waste ground, roadsides, river banks and wood borders; rare. Alien.
1. Bank of the Afon Honddu, Llanthony, *Davies:* bank of the River Wye, near Hadnock Quarries*, *Lewis.* **3.** Llangybi Road, near Llanhennock; Llanbadoc, *Miss Jenkins:* The Warrage, near Raglan; near Pontllanfraith*. **4.** Near Bigsweir Station*, for several years; wood border, Mounton*; Newchurch, *Shoolbred:* Botany Bay, Tintern*, *Charles.*

ERYSIMUM L.

Erysimum cheiranthoides L.   Treacle Mustard

Waste ground; rare. Casual.
4. Near Tintern Abbey, *Monington.*

CHEIRANTHUS L.

Cheiranthus cheiri L.   Wallflower

Old walls; occasional. Denizen.
1. Monmouth, *Charles.* Llanthony Priory*, *Mathews, Charles, Miss Frederick:* Abergavenny; Llanfoist. **3.** Usk†; Raglan†, *Clark:* Caerleon. **4.** Tintern Abbey; Chepstow Castle*†, *Clark, Hamilton, Shoolbred,!.*

ALLIARIA Scop.

Alliaria petiolata (Bieb.) Cavara & Grande   Garlic Mustard
   Sisymbrium alliaria (L.) Scop.

Hedgerows and woods; common. All districts. Native.

SISYMBRIUM L.

Sisymbrium officinale (L.) Scop.   Hedge Mustard

Hedgerows, roadsides, waste and cultivated ground; common. All districts. Native.

Sisymbrium orientale L.   Eastern Rocket

Waste ground and roadsides; rare. Casual.
1. Abergavenny; Monmouth*; Hadnock Farm, *Charles.* **2.** Rhymney Bridge*. **3.** Caerleon. **4.** Redbrook, *Charles.*

Sisymbrium altissimum L.   Tall Rocket

Waste ground and roadsides; rare. Casual.
1. Monmouth*, *Charles:* Llanfoist*. **3.** Caerleon. **5.** Alexandra Dock, Newport*, *Macqueen:* Rumney*.

CRUCIFERAE

ARABIDOPSIS (DC.) Heynh.

Arabidopsis thaliana (L.) Heynh.   Thale Cress
Sisymbrium thalianum (L.) Gay
Walls, dry banks, railway tracks, roadsides, waste and cultivated ground; common. All districts. Native.

CAMELINA Crantz

Camelina sativa (L.) Crantz   Gold of Pleasure
Cultivated ground; rare. Casual.
1. Kymin Hill, Monmouth*; *Charles*. **3.** Cefnila, *Clark*.

DESCURAINIA Webb & Berth.

Descurainia sophia (L.) Webb ex Prantl   Flixweed
Sisymbrium sophia L.
Waste ground; rare. Casual.
1. Kymin, Monmouth*, *Charles*. **4.** Tintern, *Druce*.

RESEDACEAE
RESEDA L.

Reseda luteola L.   Weld
Roadsides, waste places, railway banks, walls and quarries; locally frequent, especially on calcareous soils. Native.
1. Monmouth, *Richards:* Hadnock*; Redding's Enclosure*; Monmouth; Lady Park Wood; Dixton†, *Charles:* Abergavenny; near Wyesham.
2. Risca. **3.** Near Pontsampit†; near Beech Hill, Usk, *Clark:* Pont-hir; between Penpelleni and Little Mill; Croes Lan-y-fro, Bettws. **4.** Chepstow; St. Lawrence; Pandy Mill; Itton*; Portskewett; Caldicot; Dinham; Wynd Cliff, *Shoolbred:* Redbrook*; Whitebrook Halt, *Charles:* Ifton, *Mrs. Ellis:* Trelleck. **5.** Newport, *Hamilton,!:* Rumney*.

Reseda lutea L.   Wild Mignonette
Cultivated ground, roadsides, railway tracks and waste places; rare. Native.
1. Between Monmouth and Redbrook*; near Hadnock Quarry*, *Charles*.
2. Near Machen, *Hamilton*. **4.** Wye Valley, *Hamilton:* Redbrook; Penallt*, *Charles:* near Severn Tunnel, *Mrs. Ellis:* Caerwent Quarry, *T. G. Evans*.
5. Banks of the Usk, Newport†, *Clark:* St. Brides Wentlloog, *Hamilton:* Newport Docks, *Hamilton.!.*

Reseda alba L.   White Mignonette
Waste ground; rare. Established alien.
5. Newport Docks, 1968.!.

VIOLACEAE
VIOLA L.

Viola odorata L.　Sweet Violet

Hedge banks, open woods and railway banks; common, except in district 2. where it is rare. Native.
Var. **dumetorum** (Jord.) Rouy & Fouc. and var. **imberbis** (Leighton) Henslow are fairly common. The var. **subcarnea** (Jord). Parl. is frequent in Districts 1 and 4.

Viola hirta L.　Hairy Violet

Rocky limestone banks, and woods and hedge banks on calcareous soils; locally frequent. Native.
1. Wye Valley near Monmouth, *Hamilton:* Monmouth; Lady Park Wood, *Charles.* **2.** Abergavenny district, *Hamilton.* **4.** The Barnetts; Chepstow*; Trap Hill, Mounton*; Rogiet; Llanfihangel; St. Brides Netherwent; near St. Lawrence; cliffs east of Mounton*; between Wynd Cliff and Temple Door, St. Arvans*, *Shoolbred:* Wynd Cliff, *Ley, Shoolbred.*

×**odorata** (V. ×*permixta* Jord.)
4. Trap Hill, Mounton*; Wynd Cliff; Thornwell; Portskewett, *Shoolbred.*

Viola riviniana Reichb.　Wood Violet

Woods, hedge banks and heaths. Native.
ssp. **riviniana** Common. All districts.

ssp. **minor** (Gregory) Valentine.
In more open situations than ssp. *riviniana.*
2. Machen*. **3.** Coed Mawr, near Risca*. **4.** Wynd Cliff; Barnett Woods*, *Shoolbred.*

Viola reichenbachiana Jord. ex Bor.　Pale Wood Violet
Viola sylvestris auct.

Woods and hedge banks; common on calcareous soils of Districts 1 and 4; unrecorded elsewhere. Native.

×**riviniana** (*V.* ×*intermedia* Reichb.)
4. Wynd Cliff, *Marshall and Shoolbred.*

Viola canina L.　Dog Violet

Woods and heaths on calcareous soils; very rare. Native.
4. Wood near Tintern, *Shoolbred.*

×**riviniana**
4. Wood near Tintern, *Shoolbred.*

VIOLACEAE

Viola lactea Sm.   Smith's Violet

Rough pasture; very rare. Native.
4. By Minnetts Lane, Rogiet*, *Mrs. Ellis,!.*

Viola palustris L.   Marsh Violet

Bogs, marshy places and wet woods. Native.
ssp. **palustris.** Common throughout Districts 1 to 4, unrecorded from District 5.

Viola lutea Huds.   Mountain Pansy

Mountain pastures; very rare. Native.
4. Honddu and Grwyne Fawr Valleys, *Ley:* Hatterall Hill, *Miss Buisson.*
2. Blaenavon, *Clark.*

Viola tricolor L.   Wild Pansy

Cultivated and waste ground; rare. Native.
2. Ty-poeth, Pontypool, *T. H. Thomas*: Aberbeeg*. **3.** Castleton, *Drabble.* **4.** Portskewett, *Shoolbred.*

Viola arvensis Murr.   Field Pansy

Cultivated and waste ground; common. All districts. Native.

POLYGALACEAE
POLYGALA L.

Polygala vulgaris L.   Common Milkwort

Heaths, pastures, banks and open woods, especially on calcareous soils; frequent. Districts 1-4. Native.

Polygala serpyllifolia Hose   Heath Milkwort

Heaths, mountain pastures and moors; common. Districts 1-4. Native.

HYPERICACEAE
HYPERICUM L.

Hypericum androsaemum L.   Tutsan

Hedgerows, woods and bushy places; frequent in Districts 1, 3 and 4, rather rare in District 2, unrecorded from District 5. Native.

Hypericum hircinum L.   Goat St. John's Wort

Woods; rare. Denizen.
4. The Coombe, *Shoolbred.*

HYPERICACEAE

Hypericum perforatum L.   Perforate St. John's Wort

Roadsides, bushy places, field borders, railway banks and waste ground; frequent. All districts. Native.

Hypericum maculatum Crantz   Imperforate St. John's Wort

Roadsides, hedgebanks, wood borders and railway banks; frequent. Districts 1-4. Native.

Hypericum tetrapterum Fr.   Square-stalked St. John's Wort

Marshes, stream sides, wet roadsides and wet woods; frequent. All districts. Native.

Hypericum humifusum L.   Trailing St. John's Wort

Gravelly or sandy soils of banks, dry pastures and woodland paths; frequent. Districts 1-4. Native.

Hypericum pulchrum L.   Upright St. John's Wort

Woods, hedgerows, shady streamsides and heaths; common in Districts 2 and 3, frequent in Districts 1 and 4, unrecorded from District 5. Native.

Hypericum hirsutum L.   Hairy St. John's Wort

Woods, bushy places, roadsides and hedgerows on calcareous soils; common in Districts 1, 3 and 4; rare in Districts 2 and 5. Native.
2. Abersychan, *Freer*. **5.** Spencer Steel Works.

Hypericum montanum L.   Mountain St. John's Wort

Woods, roadsides and railway banks on limestone soils; rare and local. Native.
1. Railway bank opposite Mally Brook; near Highmeadow siding; near Hadnock Quarry*; Kymin Hill, Monmouth*; near Fiddler's Elbow; Garth Wood, near Monmouth*; railway bank opposite Dixton*; Hadnock Wood*; Lady Park Wood*, *Charles*. **4.** Wynd Cliff, *Motley, Shoolbred:* Chepstow; Mounton; Shirenewton; Llandogo, *Shoolbred:* Piercefield Wood, *Redgrove*.

Hypericum elodes L.   Bog St. John's Wort

Bogs and acid marshes; rare. Native.
2. Near Pontnewydd, *Conway:* Blaenavon; Pontypool, *Clark:* Behind Puzzle House, Bargoed; Pen-y-fan Pond, *McKenzie:* Near Foxwood, *Hamilton:* Mynydd Dimlaith*. **4.** Trelleck Bog*, *Ley, Hamilton, Charles,!:* Pen-y-fan, near Whitebrook, *Charles:* by the Virtuous Well, Trelleck.

CISTACEAE
HELIANTHEMUM Mill.
Helianthemum nummularium (L.) Mill.   Rockrose
H. chamaecistus Mill. *H. vulgare* Gaertn.

Dry banks and pastures on calcareous soils and limestone rocks; locally frequent. Native.
4. Wynd Cliff†, *Clark, Shoolbred,!:* Penmoyle†, *Clark:* Mounton; Shirenewton; Runstone; Dinham; The Minnetts; near Llanvair Discoed; near Caldicot; Crick; Llanmelin* (a form with a bright orange-coloured ring at the base of the petals), *Shoolbred:* Portskewett; Ifton, *Hamilton:* Highmoor Hill, *Mrs. Ellis:* near Carrow Hill.

CARYOPHYLLACEAE
SILENE L.
Silene vulgaris (Moench) Garcke   Bladder Campion
S. cucubalis Wibel

Hedge banks, railway banks, rough pastures, wood borders, cultivated and waste ground; locally frequent or occasional. All districts. Native. Plants with pubescent leaves are not uncommon.

ssp. **maritima** (With.) A. & D. Löve   Sea Campion

Near the coast; rare. Native.
5. About Newport, *Clark:* near Town Dock, Newport, *Hamilton.*

Silene dichotoma Ehrh.

Cultivated ground; very rare. Casual.
1. Field on Kymin Hill, Monmouth*, *Charles.*

Silene gallica L.   English Catchfly

Cultivated and waste ground, and roadsides; rare. Colonist.
1. Abergavenny, *White:* Wye Valley, Monmouth, *Hamilton:* Kymin Hill, Monmouth, *Charles:* near Pandy Station, *Rees.* **3.** Monkswood, *Clark:* garden weed, Usk Priory*, *Rickards.* **4.** Kilgwrrwg*, *Shoolbred:* Upper Redbrook*; *Charles:* Highmoor Hill, *Mrs. Ellis.*

Silene nutans L.   Catchfly

Waste ground; rare. Established alien.
5. Newport Docks*, 1944, *Macqueen!.*

Silene italica Pers.   Italian Catchfly

Waste ground; rare. Alien.
5. Newport Docks*, 1953, *Macqueen.*

CARYOPHYLLACEAE

Silene dioica (L.) Clairv.   Red Campion
Lychnis dioica L.

Hedgebanks, woods and bushy places; common. All districts. Native.

Silene alba (Mill.) E. H. L. Krause   White Campion

Hedgerows, railway banks and cultivated ground; occasional to locally common. Native.
1. Monmouth, *Charles:* Llanfoist*; Ponty-y-spig. **2.** Between Bedwas and Tre-hîr; near Gelligroes. **3.** Trostra*, *Conway:* Foxwood and Church Road, *Hamilton:* Llandegfedd; Plas Machen. **4.** Common. **5.** Between Marshfield and Castleton; Newport Docks.

LYCHNIS L.

Lychnis flos-cuculi L.   Ragged Robin

Marshes, wet meadows and streamsides; common. All districts. Native. White-flowered forms are recorded from (**1**) near Pen-y-clawdd and (**4**) Trelleck.

AGROSTEMMA L.

Agrostemma githago L.   Corn Cockle

Cultivated and waste ground, and roadsides; rare. Colonist or casual.
1. Kymin Hill, Monmouth*; Wyesham*, *Charles.* **2.** Between Gelligroes Mill and Ynysddu, *McKenzie.* **3.** Raglan, *Miss Frederick:* Croes Lan-y-fro, Bettws*. **4.** Caerwent; Dewstow, *Hamilton:* near Chepstow; Llanvair Discoed; Portskewett, *Shoolbred:* Little Llanthomas*; Old Llanishen, *R. Parry:* Magor, *Mrs. Ellis.* **5.** Alexandra Dock, Newport*; *Macqueen.*

DIANTHUS L.

Dianthus caryophyllus L.   Clove Pink

Walls; very rare and apparently extinct in the recorded localities. Alien.
4. Mathern, 1800, *anon.* on ruined walls about Tintern Abbey, *Clark.*

Dianthus armeria L.   Deptford Pink

Pastures; very rare. Native.
1. Wye Valley, near Monmouth, *Hamilton.*

SAPONARIA L.

Saponaria officinalis L.   Soapwort

Gravelly banks of rivers, by streams and waste ground; common along the rivers Rhymney, Usk, Wye and Monnow, occasional elsewhere. Districts 1-4. Denizen.

CARYOPHYLLACEAE

CERASTIUM L.

Cerastium arvense L.   Field Mouse-ear Chickweed

Waste ground and ballast; rare. Casual.
5. By the River Usk, Newport†, *Clark*.

Cerastium fontanum Baumg. ssp. **triviale** (Murb.) Jalas   Common Mouse-ear Chickweed
  C. vulgatum auct. *C. holosteoides* Fr.

Pastures, roadsides, walls and cultivated ground; very common. All districts. Native.

Cerastium glomeratum Thuill.   Broad-leaved Mouse-ear Chickweed
  C. viscosum auct.

Roadsides, pastures, walls, gravelly places and cultivated ground; common in Districts 1, 3 and 4, rare to frequent elsewhere. Native.
An apetalus form was found by Shoolbred on the sandy shore of the Severn at Mathern*.

Cerastium diffusum Pers.   Sea Mouse-ear Chickweed
  C. tetrandrum Curt.

Dry banks near the sea; rare. Native.
5. By the Severn, Sudbrook*, *Marshall & Shoolbred:* sea wall, Rumney*.

Cerastium semidecandrum L.   Little Mouse-ear Chickweed

Limestone quarries, walls and dry places; rare. Native.
3. Machen*. **4.** Portskewett*; Newchurch, *Shoolbred.*

MYOSOTON Moench

Myosoton aquaticum (L.) Moench   Water Chickweed
  Stellaria aquatica (L.) Scop.

Stream and riversides, ditches and alder swamps; locally frequent. Native.
1. Monmouth*; Osbaston*; Llantilio Crossenny*; Dixton; Grosmont*; Skenfrith*; near Mitchel Troy*; near Monmouth Cap, *Charles:* Llanfoist*. **3.** Allt-yr-yn, *McKenzie.* **4.** Wye banks, *Ley:* St. Bride's Netherwent; Brockweir, *Shoolbred:* Llandogo, *Shoolbred, Mrs. Leney:* Whitebrook*; Penallt*; near Redbrook*, *Charles.* **5.** St. Bride's Wentlloog; Nash; Goldcliff, *Hamilton:* Greenland Reen, near St. Mellons*; Magor*; Spencer Steel Works site.

STELLARIA L.

Stellaria nemorum L.   Wood Chickweed

Woods; rare and confined to the Wye Valley. Native

CARYOPHYLLACEAE

ssp. **nemorum**
1. Bank of the River Wye, Hadnock*, *Charles.* **4.** Penallt*, *Charles:* Tintern*, *Lousley.*
ssp. **glochidosperma** Murbeck
4. Llandogo*, *Ley.*

Stellaria media (L.) Vill.   Chickweed
An abundant weed of cultivated ground, roadsides and waste ground. All districts. Native.

Stellaria pallida (Dumort.) Piré   Pale Chickweed
Stellaria apetala auct.
Sandy places chiefly near the sea; rare. Native.
2. Blaen-y-cwm*. **5.** Severn banks below Mathern*; near Black Rock, Portskewett; near Severn Tunnel Junction, *Shoolbred.*

Stellaria neglecta Weihe
Streamsides, ditches, shady hedgerows, woods and moist bushy places; frequent in Districts 1, 3 and 4, rare in District 5, absent from District 2. Native.

Stellaria holostea L.   Greater Stitchwort
Hedgerows, copses and woods; common. All districts. Native.

Stellaria graminea L.   Lesser Stitchwort
Grassy roadsides, hedgebanks, heaths and bushy places; common. All districts. Native.

Stellaria alsine Grimm   Bog Stitchwort
S. uliginosa Murray
Marshes, boggy places, streamsides and springs; common, especially in upland areas. All districts. Native.

SAGINA L.

Sagina apetala Ard.   Annual Pearlwort
Wall bases, between paving stones, gravelly places, railway tracks and cultivated ground. Native.

ssp. **erecta** (Hornem.) Hermann
S. apetala auct.
Common. All districts.

ssp. **apetala**
S. ciliata Fries
1. Abergavenny, *Druce.* **2.** Rhymney Bridge*.

CARYOPHYLLACEAE

Sagina maritima Don   Sea Pearlwort

In short turf near the sea; rare. Native.

5. Path of the sea wall and bank of the River Rhymney, near Rumney*; St. Brides Wentlloog*.

Sagina procumbens L.   Procumbent Pearlwort

Wall bases, between paving stones, gravel paths, heaths, marshy places and cultivated ground; very common. All districts. Native.

(**Sagina nodosa** (L.) Fenzl. Recorded by Hamilton from Dyffryn and Langstone but probably in error since both localities are unlikely for this species.)

MINUARTIA L.

Minuartia hybrida (Vill.) Schischk.   Fine-leaved Sandwort
Arenaria tenuifolia L.

Rare. Alien.

1. On railway ballast, Hadnock*, 1944, *Charles.*

HONKENYA Ehrh.

Honkenya peploides (L.) Ehrh.   Sea Purslane
Arenaria peploides L.

Gravelly or sandy places by the Severn; rare. Native.

5. Severn coast, *Clark:* St. Bride's Wentlloog; Peterstone Wentlloog, *Hamilton.*

MOEHRINGIA L.

Moehringia trinervia (L.) Clairv.   Three-nerved Sandwort
Arenaria trinervia L.

Hedgerows and woods; common. All districts. Native.

ARENARIA L.

Arenaria serpyllifolia L.   Thyme-leaved Sandwort

Wall tops, dry banks, roadsides, quarries and cultivated ground; frequent. All districts. Native.

Arenaria leptoclados (Reichb.) Guss.   Slender Sandwort

Wall tops, quarries, colliery tips, dry waste places and cultivated ground; widespread but not common. Native.

1. Abergavenny, *Druce:* Hadnock*, *Charles:* Cwmyoy*, *Mrs. Vaughan:* near Troy Station, Monmouth. 2. Pontypool, *Bladon:* near Tredegar*. 4. Chepstow, *Shoolbred, T. G. Evans:* Trelleck, *Watkins:* Ifton*. 5. Between Castleton and Marshfield*; Newport Docks*.

76

CARYOPHYLLACEAE

SPERGULA L.

Spergula arvensis L.   Corn Spurrey

Cultivated ground and waste places; frequent to common. Districts 1-4. Native.

Both var. **arvensis** and var. **sativa** (Boenn.) Mert. & Koch are recorded.

SPERGULARIA (Pers.) J. & C. Presl

Spergularia rubra (L.) J. & C. Presl   Red Sand Spurrey

Dry gravelly places; rare. Native.
2. Near Pontypool*, *T. H. Thomas:* between Abertysswg and Tredegar; Cefn Brithdir, *McKenzie:* slag heap, Cwmbran, *Freer:* colliery tip, Crosskeys*. **4.** Near Trelleck, *Ley.*

Spergularia media (L.) C.Presl   Greater Sea Spurrey

Salt marshes; locally common. Native.
4. Common on the banks of the Severn and Wye, Chepstow*†, *Clark, Ley, Shoolbred:* St. Pierre Pill*, *Shoolbred.* **5.** Newport†, *Clark, Smith:* Undy*, *Shoolbred:* Peterstone Wentlloog*; Rumney; near Magor Pill*.

Spergularia marina (L.) Griseb.   Lesser Sea Spurrey

Salt marshes; locally common. Native.
4. Common by the Wye and Severn; Chepstow*, *Shoolbred.* **5.** Denny Island, *Mrs. M. L. Davis:* Peterstone Wentlloog*; Rumney.

(Shoolbred's records for *Spergularia rupicola* in his *Flora of Chepstow* are based upon erroneous identifications.)

ILLECEBRACEAE

SCLERANTHUS L.

Scleranthus annuus L.   Annual Knawel

Cultivated ground; common in the eastern half of the county, rare elsewhere. Native.
1. and **4.** Common. **3.** Between Castleton and Fairwater*; Llandegfedd*.

PORTULACACEAE

MONTIA L.

Montana fontana L.   Water Blinks

Swamps, and by rills and springs; common in upland areas, rare elsewhere. Native.

77

PORTULACACEAE

ssp. **amporitana** Sennen
Montia intermedia Walters
2. Cwm Carn, Abercarn*; Cwm Lickey, near Pontypool*; Cwm Lasgarn, Abersychan*. **3.** Near Malpas*; near Penheol-y-badd-Fawr, Henllys*. **4.** Trelleck Bog*, *Shoolbred, T. G. Evans.*

ssp. **variabilis** Walters
4. Near Chepstow Park*; Mounton Brook*; Tintern*; Yellow Moor*; Rogerstone Grange, St. Arvans*, *Shoolbred.*

CLAYTONIA L.
Claytonia alsinoides Sims.
Casual.
1. Dingestow*, *Gwilliam.*

AMARANTHACEAE
AMARANTHUS L.
Amaranthus albus L.
Casual.
4. Garden weed, Mathern*, *Shoolbred.*

CHENOPODIACEAE
CHENOPODIUM L.
Chenopodium bonus-henricus L.   Good King Henry
Roadsides and waste ground, especially about farm buildings; widespread but rather uncommon. Colonist.
1. Abergavenny district, *White:* near Llanthony Abbey*; Govilon, *Charles:* Llanfoist. **2.** Rhymney*; Hafodyrynys; Trefil*. **3.** Raglan district, *Miss Frederick:* canal bank between Newport and Malpas*. **4.** Near Severn Tunnel, *Hamilton:* Chepstow; between Chepstow and Crossway Green; Portskewett; Mounton; Penterry; St. Arvans*, *Shoolbred:* Mitchel Troy*, *Charles:* Carrow Hill, *Mrs. Ellis:* Magor*. **5.** Undy, *T. G. Evans.*

Chenopodium polyspermum L.   Many-seeded Goosefoot
Waste and cultivated ground; rare. Colonist.
1. Hadnock Farm*, *Charles.* **3.** Raglan district, *Miss Frederick:* Llanhennock*, *Mrs. Parris:* Caerleon; Llanddewi Fach; Llandegfedd*. **4.** Mathern, *Shoolbred.* **5.** Rumney*; Undy*.

Chenopodium vulvaria L.   Fetid Goosefoot
Waste ground; very rare. Casual.
3. Raglan district, *Miss Frederick.* **4.** Chepstow Castle, *Walford*, 1781.

CHENOPODIACEAE

Chenopodium album L.   White Goosefoot

Cultivated and waste ground; common. All districts. Native.

var. **candicans** Lam.
3. Castleton.*

var. **viridescens** St. Amans.
Common.

Chenopodium ficifolium Sm.   Fig-leaved Goosefoot

Cultivated and waste ground; rare. Colonist or casual.
4. Mathern, *H. A. Evans:* between Sudbrook and Severn Tunnel Junction; Rogiet*, *Shoolbred:* Severn Tunnel Junction*; Magor*; Undy.

Chenopodium murale L.   Nettle-leaved Goosefoot

Rare. Casual.
5. Newport Banks, *Clark.*

Chenopodium urbicum L.   Upright Goosefoot

Rare. Casual.
1. By the Wye Bridge, Monmouth, *F. A. Rogers.* **5.** Site of a manure heap, near St. Mellons.*

Chenopodium rubrum L.   Red Goosefoot

Cultivated ground, waste places and roadsides; locally frequent. Colonist.
1. Monmouth*, *More, Charles:* Chippenham*, *Charles.* **4.** Mathern*; Ifton Hill, *Shoolbred.* **5.** Between Marshfield and St. Brides Wentlloog; between Llanwern and Whitson; Goldcliff.

var. **pseudo-botryoides** H. C. Wats.
4. On dry mud, St. Pierre Pond*, *Shoolbred.*

BETA L.

Beta vulgaris L. ssp. **maritima** (L.) Thell.   Sea Beet

Shingly places by the Severn; rare. Native.
4. Chepstow, *Clark.* **5.** Newport*, *Clark, Davies:* mouth of the River Usk, *Hamilton:* Denny Island, *F. B. A. Welch.*

ATRIPLEX L.

Atriplex littoralis L.   Sea Orache

Salt marshes and waste places near the sea; rare. Native.
4. Wye banks, Chepstow, *Shoolbred.* **5.** Newport, *Melville.*

79

CHENOPODIACEAE

Atriplex patula L.   Common Orache

Waste and cultivated ground, and muddy banks of tidal rivers; common. All districts. Native.

Atriplex hastata L.   Halberd-leaved Orache
  A. deltoidea Bab.

Cultivated and waste ground, and muddy banks of tidal rivers; common. Districts 3-5. Native.

Atriplex glabriuscula Edmonst.

Muddy banks of tidal rivers; very rare. Native.
4. Bank of the River Wye, Chepstow*, *Shoolbred.*

AXYRIS L.

Axyris amarantoides L.

Casual.
4. Caerleon excavations*, 1929, *McLean.*

SUAEDA Forsk. ex Scop.

Suaeda maritima (L.) Dumort.   Annual Sea-blite

Common on salt marshes and muddy banks of tidal rivers. Districts 4 and 5. Native.

SALICORNIA L.

Salicornia dolichostachya Moss   Glasswort

Salt marshes; rare. Native.
5. Peterstone Wentlloog*.

This and the next species were abundant at Peterstone Wentlloog until the spread of *Spartina.*

Salicornia europaea L.   Upright Glasswort

Salt marshes; locally common. Native.
5. St. Pierre; Newport, *Clark, Hamilton:* between Sudbrook and Rogiet, *Shoolbred:* Magor, *Shoolbred, T. G. Evans,!:* Peterstone Wentlloog.

Salicornia ramosissima Woods   Bushy Glasswort

Salt marshes; locally common. Native.
5. Near Mathern*; between Sudbrook and Rogiet*; Undy*; Magor*, *Shoolbred:* Peterstone Wentlloog*.

PHYTOLACCACEAE

PHYTOLACCA L.

Phytolacca americana L.   Poke Berry

Garden escape

4. St. Pierre, 1925, *H. E. Salmon.*

TILIACEAE

TILIA L.

Tilia platyphyllos Scop.   Large-leaved Lime

Rocky limestone woods; rare, apparently confined to limestone cliffs. Native.

1. Lady Park Wood*, *Ley, Charles.!.* **4.** Tintern*, *T. Lewis.*

Tilia cordata Mill.   Small-leaved Lime

Limestone woods; locally frequent to locally common. Native.

1. Near Abergavenny, *C. T. Vachell:* Lady Park Wood*; Harper's Grove, near Monmouth*; Hadnock Wood*, *Charles,!.* **2.** Lasgarn Wood, Abersychan*, *Hyde.* **4.** Dinham, *Ley, Shoolbred:* Earlswood; near Llanvair Discoed; Great Barnets Woods*; Wynd Cliff*; Liveoaks; Castle Woods, Chepstow*; below Kilpale; Piercefield Woods*, *Shoolbred,!:* near Brockweir, *Marshall:* Llanmelin Camp, *Hyde:* Kites Bushes*; Mounton*; Carrow Hill*; Black Cliff Wood; Slade Wood.

Cultivated trees at Tal-y-coed are said to have been found wild in the woods there, *Hyde.*

MALVACEAE

MALVA L.

Malva moschata L.   Musk Mallow

Roadsides, hedgerows, wood borders, railway banks, and rough pastures; frequent in Districts 1, 3, and 4, rare elsewhere. Native.

1. Cwm Bychel, Llanthony; Loxey Wood, Llanthony, *Ley:* Abergavenny, *White:* between Hadnock and Monmouth; near Wyesham Signal Box, *Charles:* Llanfoist; Llanvihangel Crucorney; Lord's Grove. **2.** Near Lower Llanfoist. **3.** Near Middle Pencarn; near Allt-yr-yn; Rhiwderyn, *Hamilton:* Raglan district, *Miss Frederick:* near Castleton*; Michaelstone y Vedw; claypit, Malpas Road, Newport; between Llantarnam and Pontnewydd; near Coed-y-paen; Penpelleni; Bettws; Henllys*; Llandegfedd; Llanddewi Fach*; Llanbadoc. **4.** Mounton; Shirenewton*; Tintern; Llanvair Discoed; Dinham; Kilgwrrwg; Oak Grove, St. Arvans; Itton; between Tintern and Trelleck; near Chepstow*; Howick, *Shoolbred:* Rogiet, *Mrs. Ellis.* **5.** Dyffryn, *Hamilton.*

MALVACEAE

Malva sylvestris L.   Common Mallow

Roadsides, waste ground, rough pastures and hedgerows; Common. Districts 1, 3-5. Denizen.

Malva neglecta Wallr.   Dwarf Mallow
Malva rotundifolia auct.

Roadsides, waste ground, about farm buildings, and railway tracks; occasional. Colonist.
1. Abergavenny district, *White:* Monmouth*, *Charles.* **2.** Pontypool*, *T. H. Thomas.* **3.** Near Newport; Malpas; Rhiwderyn, *Hamilton:* Raglan district, *Miss Frederick:* Newport*. **4.** Chepstow; Tintern; Portskewett; Magor*; Rogiet*, *Shoolbred:* Upper Redbrook*, *Charles.* Bishpool. **5.** Llanwern; Severn Tunnel Junction*.

Malva pusilla Sm.   Small-flowered Mallow
Malva rotundifolia L. *nom. ambig.*

Roadsides and waste ground; rare, Casual.
4. Near Severn Tunnel Junction, *Marshall:* Rogiet*, *Shoolbred.*

LAVATERA L.

Lavatera arborea L.   Tree Mallow

Maritime rocks and banks near the sea; rare. Native.
4. Chepstow Castle, *Walford.* **5.** Denny Island, first recorded in Parkinson's *Theatrum Botanicum,* 1640 and found to be still well established there in 1932 by L. Harrison Matthews: Sudbrook, *T. G. Evans.*

ALTHAEA L.

Althaea officinalis L.   Marsh Mallow

Salt marshes, margins of reens and ditches; rare. Native.
1. Bank of the Afon Honddu, Llanthony, *Charles.* **4.** Chepstow, *Walford:* by the Wye, Tintern, *Hamilton:* between Black Rock and Chepstow, *Herb. Ludlow.* **5.** Near Severn Tunnel Junction, *Ley:* Magor; Undy; Rogiet, *Hamilton:* between Rogiet and Magor*; Caldicot Moor, *Shoolbred:* Peterstone Wentlloog*; near Magor Pill*.

LINACEAE
LINUM L.

Linum bienne Mill.   Narrow-leaved Flax
Linum angustifolium Huds.

Pastures and roadsides on calcareous soils; rare. Native.
1. Shortlands, near the top of the Kymin, Monmouth; path leading to Buckholt Wood*; Manson's Cross*, *Charles.* **3.** Llanvihangel Llantarnam.

LINACEAE

4. Ifton, *Hamilton:* Near Portskewett; Tintern; Fairfield Farm, Chepstow*, *Shoolbred:* Crossway Green, *T. G. Evans.*

Linum usitatissimum L. Flax

Roadsides, waste and cultivated ground; rare. Casual.
1. Kymin, Monmouth, *Bicheno:* Hadnock Farm, Monmouth*; Chippenham*, *Charles.* **3.** Llanddewi Fach*. **4.** Llangwm†, *Clark:* near Bradbury Farm, Chepstow, *Dr. Clarke:* Portskewett*; St. Pierre*; Tintern, *Shoolbred:* Cwmcarvan*, *Miss Davies:* Highmoor Hill, *Mrs. Ellis.* **5.** Newport Docks*, *Smith, Macqueen.*

Linum catharticum L. Purging Flax

Pastures, roadsides, quarries and old colliery tips; common especially on calcareous soils. Districts 1-4. Native.

GERANIACEAE
GERANIUM L.

Geranium pratense L. Meadow Cranesbill

Meadows, roadsides, stream and river banks, and railway banks; locally frequent. Native.
1. Tregate Bridge, St. Maughan's*; bank of the River Monnow, Monmouth Cap; bank of the River Wye, above Monmouth*; between Monmouth and Redbrook; Buckholt Wood; Skenfrith*; near Martin's Pool*; near Grape House, Newton*; near Osbaston*, *Charles:* between Llantilio Crossenny and White Castle, *Sandwith:* Abergavenny; Govilon; Llanfoist*. **3.** Banks of the River Usk, *Clark:* near Caerleon; Newbridge on Usk, *Hamilton:* Usk, *Mrs. Scholberg:* Raglan district, *Miss Frederick:* Pant y Goitre*, *Hyde:* Malpas, *Hopkinson.* **4.** Near Bigsweir Station*, *Schoolbred:* Wynd Cliff, *Miss Woodhouse:* Severn Tunnel, *Mrs. Ellis:* Near Ty-mawr, Penallt, *Beckerlegge:* The Coombe, Shirenewton, *Lannon:* Llandevaud.

Geranium sylvaticum L. Wood Cranesbill

Wooded stream banks; very rare. Native.
1. Honddu Valley, *Ley:* Grwyne Fawr Valley*, *Ley, Mrs. Paterson, Miss Vachell, Richards.*

Geranium endressi Gay.

Field borders and waste ground; rare. Alien.
1. Buckholt*, *Charles.* **4.** Upper Redbrook; Lone Lane, Pentwyn, Penallt*, *Charles:* Cuckoo Wood, Llandogo*.

83

GERANIACEAE

Geranium versicolor L.   Streaked Cranesbill

Hedge banks; very rare. Denizen.
1. Abergavenny, *Mrs. Paterson*. **4.** Plentiful for many years between St. Arvans and Pen-y-parc*, where it was known to have been established for over 60 years prior to 1920, *Shoolbred:* near Newton Lodge, *Ley:* Grey Hill, *Dr. Clarke.*

Geranium phaeum L.   Dusky Cranesbill

Hedgerows; rare. Denizen.
1. Osbaston*, *Miss Rickards.* **4.** Near Llangwm-isaf Church†; Chepstow†, *Clark:* near Pen-y-cae-mawr, Wentwood*, *Shoolbred:* **3.** Michaelstone y vedw*. **5.** Marshfield, *Miss Neale.*

Geranium sanguineum L.   Bloody Cranesbill

Woods and roadside banks on calcareous soils; rare. Native.
4. Wynd Cliff†, *Lightfoot, Gissing, Clark, Shoolbred, Hamilton,!:* near Chepstow Castle, *Lightfoot:* Piercefield Woods, *Shoolbred:* Wye Valley, near Monmouth, *Hamilton.*

First recorded by John Lightfoot in 1773.

Geranium pyrenaicum Burm. fil.   Mountain Cranesbill

Very rare. Denizen.
4. Roadside near Tintern, *Hamilton.*

Geranium columbinum L.   Long-stalked Cranesbill

Hedgebanks, railway banks, dry stony places and rough pastures; widely dispersed but not common. Native.
1. Llanddewi Ysgyryd*; Half-way House Wood, near Monmouth*; Cwmoy*, *Charles:* near Abergavenny, *Hardaker:* Pont-y-spig*. **3.** Malpas, *Hamilton:* near Michaelstone Bridge*; near Castell-y-bwch; near Bettws; near Coed-y-bedw; Penpelleni; Ynys-y-fro; between St. Mellons and Llanrumney. **4.** Chepstow Castle, *Motley, Hamilton, Shoolbred:* Ifton, *Hamilton, Mrs. Ellis:* Mounton, Dinham; Llanvihangel Rogiet; Llanmelin*; Earlswood*; Portskewett, *Shoolbred:* Llanfair Discoed, *Shoolbred, Lannon:* Mitchel Troy, *Charles.*

Geranium dissectum L.   Cut-leaved Cranesbill

Roadsides, banks, pastures, quarries, cultivated ground and waste places; common. All districts. Native.

Geranium rotundifolium L.   Round-leaved Cranesbill

Very rare. Alien.
2. Pontypool*, *T. H. Thomas.* **4.** Old Town Wall, Chepstow*, *Shoolbred.*

GERANIACEAE

Geranium molle L. Dove's-foot Cranesbill

Roadsides, pastures, walls and cultivated ground; common. All districts. Native.

var. **aequale** Bab.
3. Llandegfedd*.

Geranium pusillum L. Small-flowered Cranesbill

Waste ground, roadsides and cultivated ground; rare. Native.
1. Near Hadnock Siding; Lady Park Wood, *Charles.* **3.** Bishpool; Rhiwderyn, *Hamilton.* **4.** Chepstow; Tintern; Sudbrook*; Rogerstone Grange, St. Arvans*, *Shoolbred:* The Coombe, Shirenewton, *Lannon.* **5.** Town Dock, Newport, *Hamilton:* Burness Castle Quarry, *T. G. Evans.*

Geranium lucidum L. Shining Cranesbill

Limestone walls and quarries, and hedgebanks on calcareous soils; locally frequent. Native.
1. Cwmyoy, *Dr. Wood, Ley:* Honddu Valley, *Ley:* near Pont-y-spig, *Rep. Woolhope Club:* Monmouth, *Charles:* Abergavenny*; between Llanvihangel Crucorney and Cwmyoy*. **2.** Craig yr Hafod, near Blaenavon, *Guile,!:* near Pontnewynydd*. **3.** Raglan district, *Miss Frederick.* **4.** Chepstow*, *Hamilton, Shoolbred:* Tintern*, *Shoolbred, Riddelsdell:* Rogiet, *Mrs. Ellis:* near Ty-mawr, Penallt, *Beckerlegge:* Ifton; Magor; Wynd Cliff; The Coombe, Shirenewton; near Penallt Old Church.

Geranium robertianum L. Herb Robert

Hedgerows, woods, walls and rocky places; common. All districts. Native. White-flowered forms are recorded from several localities in District 4.

Geranium ibericum Cav.

Garden escape. Rare.
1. roadside near the school, Cwmyoy*, 1942, *Charles.*

var. **platypetalum** (Fisch. & Mey.) Boiss.
1. Abergavenny, 1924, *Druce.*

ERODIUM L'Hérit.

Erodium moschatum (L.) L'Hérit. Musk Storksbill

Walls; very rare. Native or alien.
1. Wall top, Cwmyoy, plentiful but near a cottage and bearing no sign of being native, *Ley.* **5.** Alexandra Dock, Newport, *Hamilton:* formerly grew along the top of and on the side of the sea wall, Rumney*, destroyed on the reconstruction of the wall.

GERANIACEAE

Erodium cicutarium (L.) L'Hérit.   Hemlock Storksbill

Walls, limestone quarries, cultivated ground and waste places; rare. Native or alien.
3. Newport†, *Clark:* St. Woolas and Risca Road, Newport, *Hamilton.*
4. Chepstow; Mounton; near Tintern; Portskewett*, *Shoolbred:* Ifton, *Mrs. Ellis:* Llangoven, *Miss Davis:* Rogiet, *T. G. Evans.* **5.** Newport Docks, *Hamilton, Smith:* Sudbrook, *T. G. Evans.*

(**Erodium maritimum** (L.) L'Hérit. recorded by Hamilton as having been found near Chepstow is probably an error.)

OXALIDACEAE
OXALIS L.

Oxalis acetosella L.   Wood Sorrel

Woods, hedgebanks and shaded mountain rocks; common. Districts 1-4. Native.

var. **subpurpurascens** DC.

1. Near Llanthony Abbey, *Ley:* Redding's Enclosure*, *Charles:* Craig, Monmouth, *Webb.* **4.** The Barnett's, near Chepstow*, *Shoolbred:* Duffield's Farm, Upper Redbrook, *Charles.*

Oxalis corniculata L.   Yellow Wood Sorrel

Roadsides; rare. Garden escape.
1. Hadnock Road, Monmouth, 1942, *Charles.* **5.** Rumney.

(**Oxalis stricta** L. recorded by R. Lewis as a garden weed at Mayhill, Monmouth is an error. The plant is *Oxalis corniculata* with purple coloured leaves and doubtless originally planted.)

BALSAMINACEAE
IMPATIENS L.

Impatiens noli-tangere L.   Touch-me-not

Garden escape.
4. Established outside a garden wall in Piercefield Park, near Chepstow*, 1888, *Shoolbred.*

Impatiens capensis Meerb.   Orange Balsam

Reensides; rare. Denizen.
5. By the Alexandra Dock feeder, Tredegar Park, and reens round 'The Belt', near Tredegar Park, 1909, *Hamilton:* Percoed Reen, Coedkernew, near Dyffryn*; Marshfield.

BALSAMINACEAE

Impatiens glandulifera Royle   Indian Balsam

River sides, especially where silt is deposited; locally abundant. Denizen.
1. River Usk, Abergavenny, 1927, *Miss Post:* River Trothy between Mitchel Troy and Monmouth, *Willan:* River Monnow, Skenfrith, *Grigson:* Monmouth*; Dixton*, *Charles:* River Usk, Llanfoist. **2.** River Rhymney, below Mynydd Dimlaith, *Mrs. Pinkard.* **3.** Usk*, *Smith:* Afon Lwyd, between Llantarnam and Pontnewydd*. **4.** Near Whitebrook*; Penallt*; near Redbrook*, *Charles.*

ACERACEAE
ACER L.

Acer pseudoplatanus L.   Sycamore

Woods and hedges; commonly planted but naturalized in many places throughout the county. All districts. Denizen.

Acer campestre L.   Field Maple

Woods and hedges; common, especially on calcareous soils. All districts. Native.

HIPPOCASTANACEAE
AESCULUS L.

Aesculus hippocastanum L.   Horse Chestnut

Denizen.
4. Naturalized in Piercefield Woods.

AQUIFOLIACEAE
ILEX L.

Ilex aquifolium L.   Holly

Woods and hedges; common throughout the county. All districts. Native.

CELASTRACEAE
EUONYMUS L.

Euonymus europaeus L.   Spindle-tree

Hedges and woods; common on calcareous and Old Red Sandstone soils; rare in District 2 and absent from District 5. Native.

87

RHAMNACEAE

RHAMNUS L.

Rhamnus catharticus L.   Common Buckthorn

Woods and hedges, usually on calcareous soils; rare. Native.
1. Far Hearkening Rock, Redding's Enclosure*, *Ley, Charles:* near The Hendre, *Hamilton:* Lady Park Wood, *Charles.* **2.** Glede Farm, Bedwas. **3.** Usk, *Hamilton.* **4.** Trap Hill, Mounton*; between Chepstow and Mounton, *Shoolbred:* Wye Valley; Beech Hill, *Hamilton:* Minnetts Lane, near Rogiet*; Slade Wood.

FRANGULA Mill.

Frangula alnus Mill.   Alder Buckthorn
Rhamnus frangula L.

Woods and marshy thickets; locally frequent to locally common, absent from the north of the county. Native.
2. Lower Sirhowy from Gelligroes to Pont Lawrence Station; near Newbridge, *McKenzie:* Ty-newydd, Newbridge*, *Prout:* below Mynydd Dimlaith*. **3.** Usk, *Hamilton:* Jim Crow's Wood, between Llantarnam and Pontnewydd*; Garw Wood, Pontnewydd; near Goytre; Castell-y-bwch, Henllys*; The Park, Christchurch; near Maes Arthur, Bassaleg*; Gwern Dywyll, Llanddewi Fach*. **4.** Near Tintern, *Ley, Shoolbred:* Beech Hill, *Hamilton:* The Barnetts*, near Chepstow; Wynd Cliff Woods; Dinham; Chepstow Park Wood; Itton; between Tintern and Trelleck*; Rogiet, *Shoolbred.*

LEGUMINOSAE

GENISTA L.

Genista tinctoria L.   Dyer's Greenweed

Pastures, heaths and railway banks; locally common. Native.
1. Abergavenny; near The Hendre, *Hamilton:* between Dunkard and Hilston*; Wonastow*; Graig Hill, near Cross Ash*; Nant-y-gern, near Newcastle, *Charles:* Talycoed*, *Hyde:* Seargent's Gorse, near Monmouth. **3.** Pentwyn Farm, *Hamilton:* Raglan district, *Miss Frederick:* near Pontypool*, *T. H. Thomas:* near Usk, *Lowe:* near Malpas; Llanvihangel Llantarnam; Coed Kemeys, Bettws; Pontypool Road*; near Coed-y-paen; near Castell-y-bwch, Henllys*; Gwehelog; near Coed Newydd, Trostrey; Llanddewi Fach; Llandegfedd*. **4.** Near Chepstow, *Hamilton:* Shirenewton; below Mathern; The Glyn; between Mounton and Dinham; Trelleck*; near Rhyd-y-fedw; Kilgwrrwg, *Shoolbred:* Llangwm, *Hamilton, Grimes, T. G. Evans:* Great Llan Thomas, Cwmcarvan*; Tal-y-fan, near Raglan, *Charles:* near Milton; near Caer Lieyn, Kemeys Inferior; Mitchel Troy*.

LEGUMINOSAE

Genista anglica L.   Petty Whin

Wet pastures, heaths and copses; rare. Native.
1. Near Tregare, *Miss Frederick*. **2.** Pant yr Esk; Aberbargoed, *McKenzie:* near Pentwyn-mawr, Abercarn*, *Mrs. Leney:* near Pound-y-coedcae, Aberbeeg*. **3.** Garw, *Conway:* near Pontypool, *Bladon:* Llantarnam; Llanfrechfa, *Hamilton:* near Little Creigydd, Llanddewi Fach, *Campbell:* Garw Wood, near Pontnewydd*; Llanfrechfa Lower; Llandegfedd. **4.** Near Rhyd-y-fedw*; Coed Cae, The Glyn*, *Shoolbred*.

ULEX L.

Ulex europaeus L.   Furze, Gorse

Heaths, pastures, roadsides; common in the lowlands, less so in the upland districts. All districts. Native.

Ulex gallii L.   Western Furze

Heaths and pastures; very common in the upland districts, frequent elsewhere. Districts 1-4. Native.

SAROTHAMNUS Wimm.

Sarothamnus scoparius (L.) Wimm. ex Koch   Broom
  Cytisus scoparius (L,) Link

Heaths, roadsides, quarries and railway banks; widely distributed, chiefly on calcareous and Old Red Sandstone soils. Districts 1-4. Native.

ONONIS L.

Ononis repens L.   Creeping Restharrow

Roadsides, quarries, dry pastures and railway banks, especially on basic soils; locally frequent. Native.
1. Hadnock Farm*; near Porth-y-gaeold Farm, Rockfield*; Wonastow*; Cross Ash; near Troy Station; near Iron Bridge, Monmouth*; Onen*; near Crossways, Skenfrith*, *Charles:* Llanfoist*; Tregare; near Maerdy, Grosmont. **2.** Near Cwm Lasgarn, Abersychan; between Bedwas and Trehîr; Pandy Mawr, Bedwas*; Risca*. **3.** Christchurch; Bishpool, *Hamilton:* Raglan district, *Miss Frederick:* Pontypool Road; near Cilfeigan Park; between Penpelleni and Little Mill; by the Afon Llwyd, near Llangattock; near Pant-yr-eos, Henllys; Llandegfedd. **4.** Chepstow; Mounton; Dinham; Shirenewton; Thornwell, Chepstow; Sudbrook Camp, *Shoolbred:* Pen-y-clawdd*, *Miss Davis:* Highmoor Hill; Ifton Quarries, *Mrs. Ellis:* Bishton. **5.** On ballast, Newport, *Whitwell*.

LEGUMINOSAE

Ononis spinosa L.   Spiny Restharrow

Banks and rough pastures; frequent on the alluvial soil along the River Severn, very rare elsewhere. Native.
4. Below Thornwell; Mathern, *Shoolbred:* Bigsweir*, *Charles:* quarry, Langstone. **5.** St. Pierre Pill*; Rogiet, *Shoolbred:* Peterstone Wentlloog*; near Magor Pill*.

MEDICAGO L.

Medicago falcata L.   Sickle Medick

Waste ground. Casual.
5. Alexandra Dock, Newport, *Davies.*

Medicago sativa L.   Lucerne

Waste ground and roadsides; rare. Escape from cultivation.
1. Dixton; Kymin Hill, Monmouth*, *Charles.* **2.** Brynmawr, *Miss Leake.* **3.** Near Castleton*; near Castell-y-bwch*. **4.** Portskewett, *Shoolbred:* Rogiet, *Mrs. Ellis:* near Caerwent, *T. G. Evans.*

Medicago lupulina L.   Black Medick

Roadsides, pastures, waste places and cultivated ground; very common. All districts. Native.

Medicago polymorpha L.   Toothed Medick

Waste ground; rare. Casual.
1. Kymin Hill, Monmouth*, *Charles.* **4.** Chepstow, *Shoolbred.*

Medicago arabica (L.) Huds.   Spotted Medick

Roadsides, banks and waste ground; rare. Colonist.
1. Wyesham Road, Monmouth, *Charles:* Wyesham, *Miss Rickards.* **3.** Risca Road, Newport; Bassaleg, *Hamilton.* **4.** Banks of the Wye, Chepstow*†, *Clark, Shoolbred.* **5.** Newport*, *Conway,* 1834: near Alexandra Dock, Newport, *Hamilton.*

MELILOTUS Mill.

Melilotus altissima Thuill.   Common Melilot

Waste ground, roadsides and railway banks; rare to locally frequent. Colonist.
1. Old Dixton Road, Monmouth; Redbrook, *Charles.* **3.** Banks of the River Usk, *Clark.* **4.** Tintern*, *Ley, Shoolbred:* Llanwern, *Hamilton:* Liveoaks*; Wynd Cliff, *Shoolbred:* Chepstow, *Ley, Shoolbred, T. G. Evans:* near Little Milton. **5.** Severn Tunnel, *Hamilton:* Caldicot*, *Shoolbred:* Newport Docks, *Smith,!.*

LEGUMINOSAE

Melilotus officinalis (L.) Pall.   Field Melilot
Melilotus arvensis Wallr.

Waste ground, roadsides and quarries; rare. Colonist.
1. Monmouth*, *Charles*. **4.** Redbrook*, *Charles:* Ifton*. **5.** Caldicot,
Shoolbred.

Melilotus alba Medic.   White Melilot

Roadsides and waste ground; rare. Colonist.
1. Chippenham*; Wonastow Road, near Monmouth*, *Charles*. **4.** Upper
Redbrook*, *Charles*. **5.** Railway bank, Severn Tunnel Junction, *Mrs.
Ellis:* Newport*, *Smith:* Spencer Steel Works site*.

Melilotus indica (L.) All.   Small-flowered Melilot
Casual.
5. Newport Docks*, 1968.

TRIFOLIUM L.

Trifolium ornithopodioides L.   Bird's-foot Fenugreek
Trigonella ornithopodioides (L.) DC.

Salt marshes; very rare. Native.
5. Salt marsh and path on the sea wall, Rumney*, *O. W. Richards,!.*

Trifolium pratense L.   Red Clover

Pastures, grassy roadsides and waste ground; very common. All districts.
Native.

Trifolium medium L.   Zigzag Clover

Pastures, roadsides, railway banks and wood borders; frequent throughout
most of the county except in District 5 whence it is unrecorded.

Trifolium squamosum L.   Sea Trefoil
Trifolium maritimum Huds.

Salt marshes, pastures near sea and banks of tidal rivers; rare. Native.
3. Bank of the River Usk, *Clark:* by the River Usk, Malpas*. **4.** Chep-
stow†, *Clark, Shoolbred:* below Tintern*, *Ley, Purchas, Shoolbred.*
5. Newport*, *Conway, Bowman†:* near Rumney.

Trifolium incarnatum L.   Crimson Clover

Roadside and cultivated ground; rare. Casual or relic of cultivation.
4. Meadow near Llanvair Discoed*, *Shoolbred:* Rogiet, *Mrs. Ellis.*

LEGUMINOSAE

Trifolium arvense L.   Hare's-foot Trefoil

Rare. Casual.

5. Newport Docks*, 1943, *Macqueen.*

Although a native of Britain, *Trifolium arvense* at Newport is an introduction.

Trifolium striatum L.   Soft Knotted Trefoil

Sandy or gravelly places and limestone quarries; rare. Native.
4. Old quarry, Portskewett*, *Shoolbred.* **5.** Shore of the Severn near St. Pierre Pill*, *Shoolbred:* sea wall, Rumney*.

Trifolium scabrum L.   Rough Trefoil

Gravelly places and limestone quarries; rare. Native.
4. Portskewett*, *Marshall and Shoolbred:* Ifton Quarries*, *Mrs. Ellis.*
5. Sudbrook, *T. G. Evans.*

Trifolium subterraneum L.   Subterranean Trefoil

Very rare. Native.
5. Sandy shore of the River Severn below Mathern*, *Shoolbred.*

Trifolium hybridum L.   Alsike Clover

Waste ground, roadsides and cultivated ground; frequent. All districts. Colonist or relic of cultivation.

Trifolium repens L.   White Clover

Abundant on waste ground and in grassy places throughout the county. All districts. Native.

Trifolium fragiferum L.   Strawberry-headed Trefoil

Pastures and grassy places near the sea; locally frequent. Native.
4. Banks of the River Wye, Tintern, *Ley:* tidal banks of the River Wye, Chepstow, *Shoolbred.* **5.** Bank of the Severn, Mathern*; Magor Pill*, *Shoolbred:* Caldicot level, *Mrs. Ellis:* Marshfield, *Miss Vachell:* near Rumney.

Trifolium resupinatum L.

Waste ground and roadsides; rare. Casual or established alien.
4. Between Llanvaches and Wentwood, *Shoolbred.* **5.** Newport Docks*.

Trifolium campestre Schreb.   Hop Trefoil
   Trifolium procumbens auct.

Dry fields, wall tops, roadsides, quarries and cultivated ground; frequent to common. All districts. Native.

LEGUMINOSAE

Trifolium dubium Sibth.   Lesser Yellow Trefoil

Common throughout the county in pastures, waste ground and on roadsides and wall tops. All districts. Native.

Trifolium micranthum Viv.   Least Yellow Trefoil
Trifolium filiforme L. nom. ambig.

Pastures, woodland rides, lawns, heaths and sandy grassy places; frequent to locally common. Native.
1. Llanvihangel Ystern Llewern, *Ley:* Wyesham*; The Kymin, Monmouth*; Lady Park Wood*; Pritchard's Hill*, *Charles:* Llanthony Valley*, *R. Lewis:* Llanfoist*. **2.** Mynydd Garn Clochdy, Abersychan; Mynydd Henllys, near Risca; Blorenge. **3.** Rogerstone. **4.** Tintern, *Ley:* Runston*, *Ley, Shoolbred:* Dinham*; Shepstow; The Coombe, Shirenewton*, *Shoolbred:* Ifton Quarries, *Mrs. Ellis.* **5.** Sea wall and pasture, Rumney*.

ANTHYLLIS L.

Anthyllis vulneraria L.   Kidney Vetch

Dry banks, pastures, rocky banks and railway banks on calcareous soils; rare. Native.
2. On limestone chippings of a disused railway track, below Mynydd Dimlaith. **4.** Wynd Cliff†, *Gissing, Clark, Shoolbred,!:* Earlswood, *Shoolbred, Hamilton:* The Minnetts, *Shoolbred:* Caerwent, *Mrs. Ellis:* Carrow Hill*. **5.** Sudbrook, *T. G. Evans.*

LOTUS L.

Lotus corniculatus L.   Common Bird's-foot Trefoil

Common throughout the county in grassy places. All districts. Native.

Lotus tenuis Waldst. & Kit. ex Willd.   Narrow-leaved Bird's-foot Trefoil

Rare. Established alien.
5. Waste ground, Newport Docks*, *Whitwell,!.*
First recorded in 1881; still there 1968.

Lotus uliginosus Schkuhr.   Greater Bird's-foot Trefoil

Common in marshes and wet meadows, and by streams and ditches. All districts. Native.

GALEGA L.

Galega officinalis L.   Goat's Rue

Roadsides; rare. Garden escape.
3. Bassaleg Road, near Newport, *Miss Parry:* near Langstone*. **4.** Near Milton*.

LEGUMINOSAE

ASTRAGALUS L.

Astragalus glycyphyllos L. Wild Liquorice

Wood borders, dry banks and rocky ground on calcareous soils; Rare. Native.
4. Near Dinham, *Miss Woodhall:* Norbury Common, *Dr. Clarke:* Llanvair Discoed, *Hamilton, Shoolbred:* between Dinham and Llanmellin*, *Shoolbred*:* Carrow Hill*.

ORNITHOPUS L.

Ornithopus perpusillus L. Bird's Foot

Dry gravelly places, dry pastures and banks; rare. Native.
1. Near Kymin, Monmouth, *Clark:* ascent to the earthworks near Trewyn, *Dr. Wood:* Hatterels above Pandy, *Bull:* Kymin, Monmouth*; Beaulieu Wood*, *Charles.* **2.** Little Mountain, Pontypool, *T. H. Thomas:* Abercarn*; Cwmcarn, *McKenzie:* Cefn Rhyswg, near Abercarn*. **3.** Coed-y-paen, *Conway:* Allteryn; Foxwood, *Hamilton:* Llandegfedd*, *Campbell.* **4.** Earlswood†, *Clark:* near Trelleck Bog*, *Ley, Phillips, T. G. Evans.* **5.** Alexandra Dock, Newport, *Hamilton.*

CORONILLA L.

Coronilla varia L. Crown Vetch

Rare. Denizen.
1. Bank of the River Wye above Monmouth, *Charles.* **2.** Canal bank below Chapel of Ease, between Abercarn and Newbridge*, 1923, *McKenzie.*

HIPPOCREPIS L.

Hippocrepis comosa L. Horse-shoe Vetch

Rocky limestone banks: very rare. Native.
4. Chepstow, *Clark:* near the Wynd Cliff, *Hamilton.*

ONOBRYCHIS Mill.

Onobrychis viciifolia Scop. Sainfoin

Rough pastures; rare. Denizen.
3. Pontypool*, *T. H. Thomas.* **4.** Near Wynd Cliff; Newchurch East, *Clark:* Portskewett*, *Shoolbred:* near Minnetts Lane, Rogiet, *Mrs. Ellis.*

VICIA L.

Vicia hirsuta (L.) Gray Hairy Tare

Roadsides, railway banks, pastures, bushy places, waste and cultivated ground; frequent to locally common. All districts. Native.

Vicia tetrasperma (L.) Schreb. Smooth Tare

Hedgerows, roadside banks, bushy places and cultivated ground; frequent in District 1, rare elsewhere. Native.
1. Near Monmouth, *Whitwell:* Kymin, Monmouth*; Leasebrook Lane, near Monmouth*; Osbaston*; Onen*; Llanvetherine*; between Monmouth Cap and Grosmont*; Bailey Pit Farm, near Monmouth*; Tregate Castle Farm, near St. Maughan's*; near Abergavenny, *Charles.* **3.** Alltyr-yn; Henllys; Llanhennock, *Hamilton.* **4.** Chepstow; Mounton; Shirenewton, *Shoolbred:* near Mitchel Troy*, *Charles.*

Vicia cracca L. Tufted Vetch

Hedgerows, wood borders, rough pastures and railway banks; common throughout the county. All districts. Native.

Vicia villosa Ten. ssp. **varia** (Host.) Corb.
  V. dasycarpa auct.

Casual or colonist.
3. Cultivated field, St. Mellons*, 1949.

Vicia orobus DC. Bitter Vetch

Upland pastures; very rare. Native.
1. Abergavenny district, *White.* **2.** Near Crumlin, *Clark:* Rhyswg; near Chapel of Ease, Abercarn*, *McKenzie:* Mynyddislwyn*, *Mrs. Leney:* near Cwm Lasgarn*.

Vicia sylvatica L. Wood Vetch

Woods and hedgerows; frequent on calcareous soils, very rare elsewhere. Native.
1. Kymin Hill, *Bailey:* Garth Wood; Hadnock Wood*; Buckholt Wood*; Coleford Road, Monmouth*, *Charles:* Lord's Grove. **2.** Near Crumlin, *Clark.* **4.** Near Crick; near Caerwent, *Ray:* Chepstow*, *Hamilton, Holland:* Tintern, *Shoolbred, Perman:* between Tintern and Tidenham Tunnel; The Barnetts*; Usk Road, near Pandy Mill; Tintern Road, Itton; Runston; Mounton Road, Dinham; Wynd Cliff, *Shoolbred:* near Itton, *Lowe:* St. Pierre Great Woods, *T. G. Evans.* **5.** Newport Docks*, *Macqueen.*

First recorded by John Ray in 1662.

Vicia sepium L. Bush Vetch

Hedgebanks, woods and bushy places; common. All districts. Native.

LEGUMINOSAE

Vicia sativa L.   Common Vetch

Roadsides, waste places and cultivated ground; frequent. All districts. Relic of cultivation or alien.

ssp. **angustifolia** (L.) Gaud.   Narrow-leaved Vetch

Dry pastures, banks, railway banks and roadsides; common. All districts. Native.

LATHYRUS L.

Lathyrus aphaca L.   Yellow Vetchling

Casual.
Newport Docks, *Hamilton.*

Lathyrus nissolia L.   Grass Vetchling

Pastures and railway banks; rare. Native.
1. Llanthony, *Jukes.* **2.** Machen*, *Riddelsdell.* **4.** Near Wynd Cliff, *Clark.*
5. Peterstone Wentlloog*, *C. T. and E. Vachell.*

Lathyrus pratensis L.   Meadow Vetchling

Pastures, meadows, wood borders, hedgerows and railway banks; common. All districts. Native.

Lathyrus sylvestris L.   Narrow-leaved Everlasting Pea

Woods and hedgerows on calcareous soils; frequent in District 4, rare elsewhere. Native.
3. Between Caerleon and Newport, *Bicheno, Clark.* **4.** Caerwent; Portskewett, *Hamilton:* between Dinham and Llanmelin*; Llanvair Discoed*; Penheira; Wynd Cliff; Llanvaches*, *Shoolbred:* south of Chepstow, *Fraser:* Ifton*, *Rickards:* Rogiet, *Mrs. Ellis.* **5.** Sudbrook, *Hyde, T. G. Evans:* Spencer Steel Works*.

Lathyrus montanus Bernh.   Tuberous Bitter Vetch

Open woods, copses and shady banks; fairly common and widespread except in District 2 where it is rare. Native.

GLYCINE L.

Glycine soja S. & Z.   Soya Bean

Casual.
5. Newport Docks*, *Macqueen.*

96

LEGUMINOSAE

PHASEOLUS L.

Phaseolus vulgaris L.

Casual.

5. Newport Docks, *Macqueen.*

ROSACEAE

SPIRAEA L.

Spiraea salicifolia L.   Bridewort

Hedgerows; rare. Denizen.

4. Whitebrook, 1895, *Ley:* well established about Tintern, *Webb.* Near Lady Mill, Mounton*.

FILIPENDULA Mill.

Filipendula vulgaris Moench   Dropwort
   Spiraea filipendula L.

Dry pastures and bushy places on calcareous soils; rare. Native.
4. Near Chepstow, *Clark:* near Monmouth, *Hamilton:* Minnetts Lane, near Rogiet*, *Mrs. Ellis.*

Filipendula ulmaria (L.) Maxim.   Meadowsweet
   Spiraea ulmaria L.

Moist meadows, ditches, streamsides, marshes, wet roadsides and damp woods; very common. All districts. Native.

RUBUS L.

Rubus saxatilis L.   Stone Bramble

Rocky limestone woods; locally common. Native.
4. About Chepstow, *Clark:* Wynd Cliff, *Shoolbred, Hamilton, Mrs. Ellis:* Tintern Woods, *Richards.*

Rubus ideaus L.   Raspberry

Woods, damp thickets and hedgerows; frequent throughout the greater part of the county; unrecorded from District 5. Native, but no doubt often bird sown from cultivated plants.
A variety with yellowish-white fruits is recorded by Shoolbred from Trelleck Grange.

Rubus caesius L.   Dewberry

Woods and hedgerows; frequent. All districts. Native.

97

ROSACEAE

Rubus fruticosus L. aggregate.   Bramble, Blackberry

Woods, hedgerows and bushy places; very common. All districts. Native. Many segregates of *Rubus fruticosus* L. have been recorded from Monmouthshire but recent work has shown that some of the records are based upon incorrect determinations.

The following have been verified by Mr. E. S. Edees. The list does not represent the true distribution of the Rubi in Monmouthshire since most of the work on the genus has been almost confined to the Wye Valley area.

Rubus nessensis W. Hall

4. Near Tintern*; Wentwood*, *Shoolbred:* Chepstow Park Wood*, *Riddelsdell, Edees.*

Rubus scissuss W. C. R. Wats.

4. Beacon Hill; Trelleck Common*, *Ley:* Chepstow Park Wood*; St. Arvans*, *Shoolbred.*

Rubus opacus Focke ex Bertram

4. The Narth*, *Shoolbred:* Trelleck Bog*.

Rubus affinis Weihe & Nees

4. Coed y Mynydd Common, *Ley:* Slade Wood*; near St. Brides Netherwent*, *Shoolbred.*

Rubus conjungens (Bab.) W. C. R. Wats.

3. Between St. Mellons and Llanrumney*. **4.** Shirenewton*, *Shoolbred.*

Rubus balfourianus Bloxam ex Bab.

2. Machen, *Ley.* **3.** Llanrumney*, **5.** Nash*.

Rubus halsteadensis W. C. R. Wats

4. Trelleck*, *Ley:* Beacon Hill*; Rogerstone Grange, St. Arvans*, *Shoolbred.*

Rubus dasycoccus W. C. R. Wats.

4. Beacon Hill, near Trelleck*, *Ley and Rogers, Shoolbred, Riddelsdell, Edees.*
Beacon Hill is the only station for this species in Britain apart from an unconfirmed record from Cardiganshire.

Rubus sciocharis Sudre

1. Cwmyoy, *Ley:* near Llanthony, *Edees.* **4.** Chepstow Park Wood, *Edees.*

Rubus carpinifolius Weihe & Nees

4. St. Arvans Grange*; The Glyn, Itton*; near Saw Mill, Itton*; Shirenewton*, *Shoolbred:* Tintern*, *Purchas and Ley, Shoolbred.*

Rubus nemoralis P. J. Muell.

4. Shirenewton*; Itton*, *Shoolbred.*

Rubus cambrensis W. C. R. Wats.

4. Near Bigsweir, *Ley.*

Rubus lindleianus Lees

4. Chepstow*; Shirenewton*, *Shoolbred.*

Rubus pyramidalis Kalt.

4. Wye Woods*, *Shoolbred.*

Rubus polyanthemus Lindeb.

4. Between Tintern and Trelleck*; Itton*, *Shoolbred.*

Rubus rubritinctus W. C. R. Wats.

3. Castleton*; near Rumney*. **4.** Bigsweir*; Whitebrook*, *Ley:* Trelleck Common*; Wentwood*; St. Arvans*, *Shoolbred:* near Portskcwett*, *Riddelsdell.*

Rubus acclivitatum W. C. R. Wats.

1. Cwm Bwchel, Llanthony, *Ley.* **4.** Trelleck, *Ley.*

Rubus prolongatus Boulay & Letendre

1. Wyaston Leys*, *Riddelsdell.*

Rubus cardiophyllus Muell. & Lefev.

4. Shirenewton*; near St. Arvans*; The Narth*, *Shoolbred.*

Rubus lindebergii P. J. Muell.

4. Shirenewton*, *Shoolbred.*

Rubus imbricatus Hort

4. Whitebrook Valley*, *Ley, Edees:* behind Troy House*, *Shoolbred:* Redbrook, *Watson.*

ROSACEAE

Rubus ulmifolius Schott
3. Llandegfedd*. 4. Itton*; between Tintern and Trelleck*, *Shoolbred*.
5. Peterstone Wentlloog*, *Harrison:* Rumney.*

Rubus pseudobifrons Sudre
3. Near Lan-y-nant, Llanddewi Fach*.

Rubus spengelii Weihe
4. Above Catbrook, Tintern, *Ley:* Itton*; Chepstow Park Wood*,
Shoolbred.

Rubus lentiginosus Lees.
3. Rumney*.

Rubus vestitus Weihe & Nees
1. Near Monmouth*, *Riddelsdell.* 4. Near Chepstow, *Babington:* Lower
Earlswood*, *Shoolbred.*

Rubus orthoclados A. Ley
4. Beacon Hill*, *Ley, Edees.*
First found here.

Rubus echinatus Lindl.
1. Llangattock Vibon Avel*, *Riddelsdell.* 4. Bigsweir, *Ley:* Barnett
Woods*; St. Arvans Grange, *Shoolbred.*

Rubus echinatoides (Rogers) Sudre
4. Trelleck, *Ley.*

Rubus rudis Weihe & Nees
1. Redding's Enclosure*, *Ley.*

Rubus foliosus Weihe & Nees
4. Between Tintern and Trelleck*, *F. A. Rogers:* Bigsweir*; Wentwood*,
Shoolbred: Chepstow Park Wood*, *Riddelsdell.*

Rubus rubristylus W. C. R. Wats.
3. Castleton*. 4. Wentwood*, *Shoolbred.*

100

Rubus cavatifolius P. J. Muell.

4. Beacon Hill, *Ley and Linton, Edees:* Chepstow Park Wood*, *Riddelsdell, Edees.*

Beacon Hill is the *locus classicus.*

Rubus pallidus Weihe & Nees

2. Daren Road, Risca*. **3.** Rumney*.

Rubus acutifrons A. Ley

1. Llangattock Vibon Avel; Buckholt Woods, *Ley.* **4.** Chepstow Park Wood*, *Shoolbred, Edees.*

Rubus scaber Weihe & Nees

4. Tintern*, *Shoolbred.*

Rubus longithirsiger Lees ex Bab.

4. Between Tintern and Trelleck*, *Shoolbred.*

Rubus rufescens Muell. & Lefev.

3. The Park, Christchurch*. **4.** Chepstow*, *Shoolbred.*

Rubus purchasianus (Rogers) Rogers

3. Llangattock Vibon Avel*, *Riddelsdell.* **4.** Troy Wood*, *Riddelsdell.*

Rubus retrodentatus Muell. & Lefev.

4. Near Devauden, *Ley.*

Rubus euanthinus W. C. R. Wats.

1. Redding's Enclosure, *Ley.*

Rubus angusticuspis Sudre

3. Coed y Gatlas, Llangattock nigh Usk*. **4.** Shirenewton*, *Ley and Shoolbred:* Beacon Hill*, *Shoolbred, Edees:* Chepstow Park Wood, *Riddelsdell, Edees.*

Rubus dasyphyllus (Rogers) Rogers

4. Itton; The Narth*, *Shoolbred.*

Rubus pallidisetus Sudre

1. Near Monmouth*, *Riddlesdell.*

ROSACEAE

Rubus elegans P. J. Muell.

4. Beacon Hill*, *Ley and Linton:* Chepstow Park Wood; near Parkhurst Rocks, Trelleck*; near Cadira Beeches, Wentwood*; between Trelleck and Tintern*, *Shoolbred.*

POTENTILLA L.

Potentilla palustris (L.) Scop.   Marsh Cinquefoil
Comarum palustre L.

Marshes; very rare. Native.
1. Llanfoist. **2.** The Blorenge, *Hamilton.*

Potentilla sterilis (L.) Garcke   Barren Strawberry

Hedgebanks and wood borders; common. All districts. Native.

Potentilla anserina L.   Silverweed

Roadsides, river banks and cultivated ground; common. All districts. Native.

(**Potentilla argentea** L. recorded by Hamilton from the Abergavenny district is almost certainly an error.)

Potentilla norvegica L.   Norwegian Cinquefoil

Casual.
4. Roadside, Redbrook*, *Charles.*

Potentilla hirta L. var. **pedata** Koch.

Casual.
4. Wye bank, near Tintern Abbey, *Bailey.*

(**Potentilla tabernaemontani** Aschers. recorded by Conway in the *New Botanists' Guide*, 1837, is doubtless an error, as is the record in White's *Guide to Abergavenny* 1886).

Potentilla erecta (L.) Räusch.   Tormentil

Banks, pastures, open woods, heaths and moorlands; common, especially in the upland districts, absent from district 5. Native.
× **reptans** (*P.* × *italica* Lehm.)
1. Upper part of the Honddu Valley, *Ley.* **4.** Between Wynd Cliff and Tintern, *Ley.* The Glyn, *Shoolbred.*

Potentilla anglica Laichard.   Trailing Tormentil
Potentilla procumbens Sibth.

Banks, heaths, pastures and open woods; frequent. All districts. Native.

ROSACEAE

× **erecta** (*P.* × *suberecta* Zimmet.)
4. Near Devauden, *Ley, Shoolbred:* The Barnetts, Chepstow; Fair Oak, Tintern*; Chepstow Park,* *Shoolbred.*

× **reptans.** (*P.* × *mixta* Nolte ex Reichb.)
1. Near Cwmyoy; Pandy, *Ley.* **4.** Chepstow; Tintern, *Ley:* Mounton; below Chepstow Park; Llanmelin; Trelenny, *Shoolbred.*

Potentilla reptans L. Creeping Cinquefoil
Roadsides, railway banks, pastures and old walls; common. All districts. Native.

FRAGARIA L.

Fragaria vesca L. Wild Strawberry
Woods and hedgerows; common. All districts. Native.

Fragaria moschata Duchesne   Hautbois Strawberry
Hedgerows and woods; rare. Denizen.
1. Near Buckholt, 1880, *Ley:* Buckholt Wood; near Wyesham signal box*, *Charles.* **4.** Usk Road, Mounton; Shirenewton Hill; between Pont-y-Saison and the Pantau*, *Shoolbred:* near Kites Bushes, Shirenewton*.

Fragaria × **ananassa** Duchesne   Strawberry
Naturalized garden escape.
2. Railway bank, Abersychan, 1923*.

GEUM L.

Geum urbanum L. Herb Bennet
Hedgerows, woods and bushy places; common. All districts. Native.

var. **grandiflorum** Schur.
4. Between St. Arvans and Penterry*, *Shoolbred.*

Geum rivale L. Water Avens
Damp woods and shady stream banks; locally frequent. Native.
1. Honddu Valley, *Mrs. Ley:* Grwyne Fawr Valley, *Ley:* Tarren yr Esgob*, *R. Lewis,!:* between Abergavenny and Triley Bridge*; between Great Triley and Crowfield; between Abergavenny and Llantilio Pertholey*. **2.** Varteg†, *Clark:* Abersychan*, *T. H. Thomas,!:* Rhymney Valley, *Hamilton:* Lasgarn Wood*. **4.** The Glyn, Itton, *Hamilton:* near Rhyd-y-fedw, Itton, *Shoolbred, T. G. Evans:* Shirenewton*, *Hamilton, Shoolbred.*

103

ROSACEAE

× **urbanum** (*G.* × *intermedium* Ehrh.)
1. Between Abergavenny and Llantilio Pertholey*; between Great Triley and Crowfield.

AGRIMONIA L.

Agrimonia eupatoria L.   Agrimony
Pastures, roadsides, hedgebanks and wood borders; common. All districts. Native.

Agrimonia procera Wallr.   Fragrant Agrimony
A. odorata auct. non Mill.
Grassy roadsides; rare. Native.
1. Between Llanddewi Ysgyryd and Llanvetherine, *Sandwith.* **3.** Between Llanrumney and St. Mellons*. **4.** Near Tintern, *Lieut. Fraser.*

ALCHEMILLA L.

Alchemilla vulgaris L. *sensu lato*   Lady's Mantle
Pastures, hedgebanks, grassy roadsides, woodland paths and river banks; frequent. Districts 1-4. Native.

First recorded in Parkinson's *Theatrum Botanicum*, 1640.

The following segregates of *Alchemilla vulgaris* are recorded.

Alchemilla vestita (Bus.) Raunk.
A. pseudo-minor Wilm.; *A. minor* auct. non Huds.
Frequent. Districts 1, 3 and 4.

Alchemilla filicaulis Buser
Very rare.
1. On the first Daren in The Honddu Valley, *Ley.*

Alchemilla xanthochlora Rothm.
A. pratensis auct.
1. Near Oldcastle, *Bishop:* between Llanthony and Capel-y-ffin*, *Charles,!.* Tarren yr Esgob*. **2.** Wood above Risca, *Bicheno:* Trevethin*, *T. H. Thomas:* Cefn Rhwswg, Abercarn*; Lasgarn Wood, Abersychan*.

Alchemilla glabra Neyg.
A. alpestris auct.
1. Llanthony district, *Ley:* near Llanthony, *R. Lewis.* **2.** Rhymney Bridge; Trefil*. **4.** Near Tintern, *Bailey.*

ROSACEAE

APHANES L.

Aphanes arvensis L.   Parsley Piert
Alchemilla arvensis (L.) Scop.

Cultivated ground, quarries, banks and wall tops; common. Native. Districts 1-4.

Aphanes microcarpa. (Boiss. & Reut.) Rothm.
Rare. Native.
2. Moorland bank, Craig yr Hafod, near Blaenavon*.

SANGUISORBA L.

Sanguisorba officinalis L.   Great Burnet
Poterium officinale (L.) A. Gray
Damp meadows, marshes and streamsides; rare. Native.
1. Near Abergavenny, *Bicheno:* between the Ysgyryd Fawr and the Sugar Loaf, in considerable abundance, *E. Lees:* Black Mountains, *Bull.* **2.** Near Pontypool, *Poole:* Bedwas, *Hamilton:* above Newbridge, *McKenzie:* near Pentwyn mawr, Abercarn, *Mrs. Leney:* Rhymney Bridge*; Cwm Lasgarn, Abersychan*. **4.** Wye Valley, near Tintern, *Hamilton:* near Llandogo, *Mrs. Leney.*

Sanguisorba minor Scop. ssp. **minor**   Salad Burnet
Poterium sanguisorba L.

Banks, roadsides and pastures on calcareous soils; locally common. Native.
1. Near Monmouth, *Hamilton, Charles.* **2.** Craig yr Hafod, near Blaenavon, *Guile.* **3.** Llanerthill Farm, Llandenny*, *Miss Harrhy.* **4.** Common.

ssp. **muricata** Briq.   Prickly Burnet

Poterium polygamum Waldst. & Kit.
Roadsides and railway banks; rare. Alien.
4. Castle Wood, Chepstow*, *Shoolbred.* **5.** Rumney*.

ROSA L.

Rosa arvensis L.   Field Rose
Hedges, thickets and woods; common. Native. All districts.

Rosa pimpinellifolia L.   Burnet Rose
R. spinosissima auct.
Very rare. Native.
1. Honddu Valley, *Ley.*

ROSACEAE

Rosa stylosa Desv.　Close-styled Rose

Hedges and woods; locally frequent. Native.

4. Near Llanmelin; Shirenewton; Llanvair Discoed, *Shoolbred*.

var. **systyla** (Bast.) Baker

4. Wyes Wood, near Tintern*; west of Trelleck Road, near Tintern; near Coppice; Howick*; Llanvair Discoed; near Dinham*; Itton Road, near Chepstow*; near The Innage, Mathern, *Shoolbred:* Kilpale*, *Ley:* Shirenewton, *Miss Vachell*.

Rosa canina L.　Dog Rose

Hedges, woods, and thickets; common. All districts. Native.

Rosa dumalis Bechst.　Glaucous Rose

R. glauca Vill. ex Lois.

Hedges; very rare. Native.

1. Cwmyoy*, *Ley*. **5.** Marshfield*.

Rosa dumetorum Thuill.　Thicket Rose

Hedges and thickets; frequent to common. All districts. Native.

Rosa obtusifolia Desv.　Felt-leaved Rose

Hedges; rare. Native.

1. Near Llangattock; Llanvihangel Crucorney. *Ley:* **4.** Pont y Saison, Wentwood; Llanfair Discoed*; near The Innage, Mathern*; Barnetts Farm, near Mounton; between Usk and Pen-y-cae-mawr*, *Shoolbred*.

Rosa tomentosa Sm.　Downy Rose

Hedges, thickets and wood borders; locally frequent. Native.

1. Near Llanthony Abbey; Gaer Hill, *Ley:* Abergavenny; Monmouth, *Hamilton*. **2.** Near Maes-y-cwmmer. **3.** Between Llantarnam and Pont-newydd*. **4.** Between Usk and Pen-y-cae-mawr*; near Pont y Saison, Wentwood; The Coombe, Shirenewton*; Whitebrook*; *Shoolbred*.

Rosa sherardi Davies　Sherard's Rose

Hedges and mountain cliffs; rare. Native.

1. Cwmyoy*, *Ley:* Tarren yr Esgob*; Staunton Road, Monmouth*; *R. Lewis*. **4.** Near Pont y Saison, Wentwood*; near Tintern*, *Shoolbred*.

Rosa villosa L.　Soft-leaved Rose

Hedges and wood borders; rare. Native.

1. Llanelen, *E. Lees:* Llanthony Valley, *Ley*. **2.** Near Maes-y-cwmmer*. **3.** Near Usk, *Reader*. **4.** Near Tintern; Trelleck; near Pont y Saison, *Shoolbred*.

Rosa rubiginosa L.   Sweetbriar

Hedges; rare. Native.
3. Llangeview, *Clark.* **4.** Llanwern; near Caerwent, *Hamilton:* near Tintern, *Shoolbred.*

Rosa micrantha Borrer ex Sm.   Small-flowered Rose

Hedges and thickets; rare. Native.
1. Llangattock Vibon Avel, *Ley:* Lady Park Wood*, *R. Lewis,!.* **4.** Dinham; Pant y Cossin, Shirenewton; Runstone, *Shoolbred.* **5.** By the Severn below Mathern, *Shoolbred.*

PRUNUS L.

Prunus spinosa L.   Blackthorn

Hedges, thickets and wood borders; common. Native. All districts.

Prunus domestica L. ssp. **domestica**   Plum

Hedges and wood borders; rare, but probably less so than the records suggest. Denizen.
1. Newton, near Monmouth*, *Charles:* near Maes-y-ffin, Honddu Valley*; Oldcastle*. **2.** Abercarn*, *Prout.* **3.** Bassaleg Road, near Castleton*; St. Mellons*. **4.** Mounton; between Mounton Road and the Rifle Butts, Chepstow, *Shoolbred:* Tintern, *Webb.* **5.** Peterstone Wentlloog*.

ssp. **insititia** (L.) C. K. Schneid.   Bullace

Hedges and wood borders; rare. Native.
1. Kymin*; Coleford Road, Monmouth*, *Charles.* **2.** Near Llanllywel, *Clark.* **4.** Castle Woods, Chepstow*; Tintern; Mounton*; Wynd Cliff; rocks south of Chepstow Station; Piercefield Woods*, *Shoolbred.*

Prunus avium (L.) L.   Gean

Woods and hedgerows; locally common in District 1, rare in District 2, and common in Districts 3 and 4. Native.
2. Near Twyn-gwyn Dingle, Pontypool*, *T. H. Thomas:* Machen, *Hamilton:* near Lower Llanfoist.
The localities in the other districts are too numerous for detailed reference.

Prunus cerasus L.   Wild Cherry

Woods, thickets and hedgerows; rare. Denizen.
1. Hadnock Wood*; Garth Wood, *Charles.* **3.** Woods about Usk†, *Clark:* near The Coldra, Christchurch, *Hamilton.* **4.** Near Tintern; The Barnetts, near Chepstow; Barbadoes Hill*; Penterry; The Minnetts Wood; wood below Kilpale*, *Shoolbred.*

ROSACEAE

Prunus padus L.   Bird Cherry

Hedgerows and thickets; frequent in the north of the county, rare elsewhere. Native.
1. Grwyne Fawr Valley, *Ley, Richards:* Honddu Valley*, *Ley:* near Pont-y-spig, *Trans. Woolhope Club:* near Cwmyoy, *Dr. Wood,!:* near Triley Bridge*; near Lower Station, Llanvihangel Crucorney*. **2.** South of Ystrad Mynach*, *Corbett:* near Gelligroes Mill, *McKenzie:* Craig yr Hafod, near Blaenavon; Lasgarn Wood, *Guile:* near Ynys-ddu*. **3.** Garcoed Wood, near Usk†, *Clark:* Glebe Farm, Bedwas*. **4.** Wynd Cliff, *Trans. Woolhope Club.*

Hamilton in his *Flora of Monmouthshire* gives the following localities: St. Julians, Chepstow, Machen and canal bank, Malpas. Confirmation is desirable since most of these localities are also given for *P. avium.*

Prunus laurocerasus L.   Cherry Laurel

Denizen.
4. Naturalized in Piercefield Woods.

COTONEASTER Medic.

Cotoneaster simonsii Baker

Denizen.
4. Lower Martridge Wood, St. Arvans*, 1931.

Cotoneaster microphylla Wall. ex Lindl.

Walls and rocks; rare.   Denizen.
3. Machen Church wall. **4.** meadow opposite Beaucliff, Tidenham (in Monmouthshire), *Shoolbred:* Cuckoo Wood, Llandogo*.

CRATAEGUS L.

Crataegus laevigata (Poir.) DC.   Midland Hawthorn
   C. oxyacanthoides Thuill.
Hedgerows; rare. Denizen.
3. Llandegfedd*. **4.** Bishton*.

Crataegus monogyna Jacq.   Hawthorn

Hedgerows, woods and hillsides; common. All districts. Native.

SORBUS L.

Sorbus aucuparia L.   Mountain Ash
Woods, hillsides and mountain cliffs; common, especially in upland areas. Absent from District 5. Native.

(**Sorbus domestica** L. recorded by Clark from a wood near Chepstow Castle is doubtless an error for *S. torminalis.*)

Sorbus anglica Hedlund   English Whitebeam

Limestone woods; rare. Native.
4. Wynd Cliff Woods*; Piercefield Woods*, *Shoolbred*.

Sorbus aria (L.) Crantz   Whitebeam

Woods, chiefly on calcareous soils; locally frequent. Native.
1. Lady Park Wood; Redding's Enclosure*; near The Far Hearkening Rock*, *Charles*. **4.** Tintern, *E. Lees, Hort:* two miles west of Chepstow, *Hort:* near Itton, *Clark:* Wynd Cliff*, *W. H. Purchas, Shoolbred:* between Chepstow and Shirenewton; Mounton; Dinham; Barnett Woods, *Shoolbred:* Llanmelin Camp*, *Hyde:* between Tintern and Wynd Cliff*, *Richards,!:* Piercefield Wood, *Shoolbred,!*.

× **torminalis** (*S.* × *vagensis* Wilmott)
4. Wynd Cliff; east side of Mounton Valley, *Shoolbred*.

Sorbus eminens E. F. Warburg

Rocky limestone woods; very rare. Native.
4. Near Tintern, *Shoolbred*.

Sorbus porrigentiformis E. F. Warburg
  S. porrigens auct.

Mountain cliffs; rare. Native.
1. Tarren yr Esgob; Honddu Valley*, *Ley, Charles,!:* Pen-y-wyrlod, Honddu Valley, *Charles*.

The Monmouthshire records for *Sorbus rupicola* Hedlund are errors for this species.

Sorbus torminalis (L.) Crantz.   Wild Service Tree

Woods, chiefly on the Carboniferous Limestone; locally frequent. Native.
1. By the River Wye above Monmouth, *Hamilton:* Half-way House, near Monmouth*; Onen*; Redding's Enclosure*; Lady Park Wood, *Charles,!:* Priory Grove Wood, near Monmouth*, *Lewis*. **3.** Wooded cliff by the Usk, Llancayo, *Miss Frederick*. **4.** Chepstow*; Piercefield*; Wynd Cliff; Tintern; Mounton; The Barnetts*; Shirenewton, *Shoolbred*.

PYRUS L.

Pyrus pyraster Burgsd.   Pear
  P. communis auct.

Hedgerows and thickets; rare. Denizen.
1. Between Monmouth and Drewen Cottages*. **3.** Pont-hir, *Hamilton*. **4.** About Chepstow, *Ley, Shoolbred, Hamilton*. **5.** Caldicot Level, *T. G. Evans*.

ROSACEAE

Pyrus cordata Desv.   Heart-leaved Pear

Hedgerows; very rare. Denizen.
1. Close to the River Wye, Dixton Parish, 1910, *Riddelsdell:* by the River Wye, Chapel Farm, near Wyaston Leys, *Ley.*
Riddelsdell's and Ley's records probably refer to the same locality.

MALUS Mill.

Malus sylvestris Mill.   Crab Apple
   Pyrus malus L.

Hedgerows, copses and wood borders.

ssp. **sylvestris**
Common in all districts. Native.

ssp. **mitis** (Wallr.) Mansfeld
Frequent in Districts 1, 3 and 4. Denizen.

CRASSULACEAE

SEDUM L.

Sedum telephium L. sensu lato   Orpine

Hedgebanks, walls and rocky places; locally frequent. Native.
1. Abergavenny district, *White:* between Hadnock Quarry and The Slaughter; Monmouth; *Charles.* **4.** Shirenewton, *Clark, Shoolbred:* Gwernesney, *Clark:* Kilgwrrwg, *Shoolbred.*

ssp. **fabaria** (Koch) Kirschlager

1. Between Onen and Penrhos*, *Charles.* **3.** Near Penpergwm*, *T. L. Williams:* below Pen-y-wain Farm, Llanddewi Fach*, *Campbell:* Trostra Lane, Panteg*, *Miss E. A. Jenkins:* near Castell-y-bwch; near Coed-y-bedw; near Monkswood; Penpelleni; between Coed Mawr and Llwch, Goytre; near Michaelstone y Vedw*. **4.** Barbadoes Wood; Tintern; near Chepstow Park; near St. Arvans; Mynydd-bach, Shirenewton*; *Shoolbred:* near Trelleck*, *Shoolbred, Amhurst:* Llanmelin*, *Miss Cooke.*

Sedum spurium M.Bieb.

Wall tops; rare. Naturalized garden escape.
4. Near Chepstow, *Shoolbred:* Newchurch West, *T. G. Evans:* Cleddon*; near Trelleck.

(*Sedum anglicum* Huds. recorded by Hamilton from between Wynd Cliff and Tintern is no doubt an error.)

Sedum album L.   White Stonecrop

Wall tops and rocky banks; rare. Denizen.
1. Buckholt*; between Monmouth and Staunton*; near Fiddler's Elbow, Monmouth*; Grosmont*, *Charles:* between Monmouth and Watery Lane. **3.** Coed-y-paen*. **4.** Tintern,!; Porth Caseg; Kilgwrrwg, *Shoolbred:* Trelleck, *Charles:* Llanvair Discoed, *Mrs. Ellis.*

Sedum acre L.   Biting Stonecrop

Walls, rocky banks, railway tracks and cuttings; common. Native. All districts.

Sedum forsteranum Sm.   Rock Stonecrop

Walls, and limestone rocks and banks; rare to locally frequent. Native.

ssp. **forsteranum**
1. Near Hadnock Quarries; opposite the Bibblings*; Wyesham*, *Charles.* **4.** Chepstow, *Clark:* Pwllmeyric Hill; Llandogo; Trelleck Grange; Wynd Cliff, *Shoolbred:* Newmills, Penallt, *Charles:* near Gaer Hill, *T. G. Evans:* Trelleck; Cleddon.

ssp. **elegans** (Lejeune) E. F. Warburg
S. rupestre auct.
1. St. Maughan's, *Charles.* **3.** Raglan; Caerleon*. **4.** Chepstow†, *Lightfoot, F. A. Lees, Clark, Shoolbred:* Wynd Cliff, *Motley, Marshall, Shoolbred:* Botany Bay, Tintern, *Perman:* Tintern Abbey, *Ley.*

First recorded by John Lightfoot in 1773.

Sedum reflexum L.   Reflexed Stonecrop
Walls and rocky banks; rare. Denizen.
4. Upper Redbrook, *Charles:* Tintern Abbey, *Miss Vachell:* Penallt*, *Charles,!:* Trelleck, *T. G. Evans:* Llanvair Discoed*; Magor.

SEMPERVIVUM L.

Sempervivum tectorum L.   Houseleek

Walls and roofs; rare. Alien.
2. Blaenavon, *Hamilton.* **3.** Near Goytre; near Ebbw Bridge; Dos Road, Newport, *Hamilton:* Raglan District, *Miss Frederick:* between Ponthir and Caerleon; White House, Llanddewi Fach.

UMBILICUS DC.

Umbilicus rupestris (Salisb.) Dandy   Wall Pennywort
Cotyledon umbilicus-veneris auct.

Walls and limestone rocks; common in Districts 1, 3 and 4; rare in District 2; unrecorded from District 5. Native.

SAXIFRAGACEAE

SAXIFRAGA L.

Saxifraga tridactylites L.   Rue-leaved Saxifrage

Wall tops and limestone rocks; locally frequent. Native.
1. Monmouth district, *Hamilton:* Llanthony*, *Ley, Charles:* Troy Farm, Monmouth, *Charles:* Llanvihangel Crucorney*; between Pandy and Old-castle*. **3.** Trostra*, *Conway:* Plas Machen*; Raglan. **4.** Chepstow, *Clark, Shoolbred:* Tintern*, *Shoolbred:* Penallt*, *Charles:* St. Pierre, *T. G. Evans:* near Penhow; Magor*; Trelleck*.

Saxifraga granulata L.   Meadow Saxifrage

Pastures and shady stream banks; rare. Native.
1. Abergavenny district, *White:* near Maerdy, Grosmont *Clark:* brook-side rocks by the Grwyne Fawr, *Ley:* near Hadnock Quarry*, *Charles:* near Triley Bridge*. **3.** Llanbadoc; near Rhadyr Mill, *Clark:* Tredunnock, *Hamilton:* between Usk and Trostrey*, *Rickards:* Raglan district, *Miss Frederick.*

Saxifraga hypnoides L.   Mossy Saxifrage

Limestone rocks, and mountain cliffs and screes; locally common. Native.
1. About Llanthony Abbey, *Ball, Hamilton, Charles:* Tarens skirting the Ffwddog, *Ley:* Chwarel y Fan*; Tarren yr Esgob*, *Charles,!.* **4.** Wynd Cliff*, *Ley, Shoolbred,!.*

Hamilton (*Flora of Monmouthshire*) gives Chepstow and Portskewett. The first probably refers to Wynd Cliff, but Portskewett seems an unlikely locality for *Saxifraga hypnoides* unless it was a garden escape.

CHRYSOSPLENIUM L.

Chrysosplenium oppositifolium L.   Opposite-leaved Golden Saxifrage

Streamsides, spring heads, water flushes, damp woods and wet mountain rocks; common. Native. Districts 1-4.

First recorded 1640: Parkinson, *Theatrum botanicum*, 425, ' . . . at Chepstow, . . .'

Chrysosplenium alternifolium   Alternate-leaved Golden Saxifrage

Sandy streamsides and damp woods; rare in Districts 1-3, locally common in District 4. Native.
1. Honddu Valley, *Bull:* near Capel-y-ffin; Grwyne Fawr Valley, *Ley:* Hadnock, *Charles:* south flank of Mynydd Pen-y-fal, *Hardaker:* near Llantilio Pertholey*; Triley Bridge*. **2.** Pontypool, *E. Lees:* **3.** Glascoed†, *Clark.* **4.** Tintern, *A. G. Purchas:* Llangwm†, *Clark:* near Tintern*,

SAXIFRAGACEAE

Miss Roper: Catbrook; Mounton*; near Pandy Mill, Itton; Llandogo Glen*; Kilgwrrwg, *Shoolbred:* Upper Redbrook*; by the River Wye, under Penallt*, *Charles:* upper parts of Cwmcarvan Brook and the upper parts of the brook between Ty-mawr, Penallt and Whitebrook, *Beckerlegge:* Black Cliff, near Tintern*.

GROSSULARIACEAE

RIBES L.

Ribes rubrum L. Red Currant
Ribes sylvestre (Lam.) Mert. & Koch

River banks, streamsides, hedgerows and wood borders; occasional. Denizen.
1. Near Monmouth, *Hamilton:* Dingestow*; between Mallybrook and Hadnock*, *Charles:* near Great Goytre, Grosmont. **2.** Lower Machen, *Hamilton.* **3.** Pant-yr-eos, Gwehelog; Mescoed Bach, Bettws*; between St. Mellons and Michaelstone y Vedw*. **4.** Llandogo; Wynd Cliff Woods; Mounton; Dinham, *Shoolbred:* Whitebrook Valley, *Shoolbred, Charles:* between Tintern and Trelleck Grange*.

Ribes nigrum L. Black Currant

Hedgerows, wood borders, river banks and limestone cliffs; rare. Denizen.
1. Plantation above the Garth, Monmouth*, *Charles:* by the River Usk, Llanfoist. **3.** Monkswood. **4.** Tintern; Mounton; Limestone cliffs east of Mounton; Pont-y-Saison*, *Shoolbred.* **5.** Near Ebbw Bridge, *Hamilton.*

Ribes alpinum L. Alpine Currant

Denizen.
1. Hedge near Llanthony Abbey, with a bush of Snowberry, *Ley.*

Ribes uva-crispa L. Gooseberry
Ribes grossularia L.

Hedgerows, wood borders and streamsides; occasional. Denizen.
1. Monmouth†, *Clark:* Redding's Enclosure*; Lady Park Wood; Halfway House Wood; Hadnock Wood; plantation, Coleford Road, Monmouth, *Charles:* Devauden, *Smith:* Abergavenny. **2.** Lasgarn Wood, Abersychan; near Llanfoist. **3.** Cilfeigan, *Conway:* Pontnewynydd; between Castleton and Cefn-llogell; near Coed-y-paen; near Cilfeigan Park*; Usk*; near Coed y Llyn, Llanover; near Pounds Hill, Newport*. **4.** Mounton; Wynd Cliff Woods; Usk Road, near Shirenewton, *Shoolbred:* near Glyn Farm, Penallt*, *Charles:* Trelenny, *T. G. Evans:* near Five Lanes; Piercefield Wood; between Llanmartin and Wilcrick; between Tintern and Trelleck Grange*; Cleddon*; near Penallt Old Church. **5.** Near Ebbw Bridge, *Hamilton.*

DROSERACEAE
DROSERA L.
Drosera rotundifolia L. Round-leaved Sundew

Bogs; rare. Native.
1. Mynydd Pen-y-fal (Sugar Loaf Mountain)†, *Clark:* Honddu and Grwyne Fawr Valleys; Ffwddog, *Ley:* near Pont-y-spig*. **2.** Abersychan†, *Clark:* Mynydd Garn-fawr, near Blaenavon, *Hamilton:* Trefil. **3.** Near Castell-y-bwch, Henllys*, *Sculthorpe,!.* **4.** Trelleck Bog*, *Hamilton, Shoolbred, Charles,!:* Barbadoes Hill; The Narth, *Shoolbred:* Whitelye Common.

LYTHRACEAE
LYTHRUM L.
Lythrum salicaria L. Purple Loosestrife

Marshes, streamsides and reens; locally frequent. Native.
1. Abergavenny districts, *White:* near Dingestow, *Clark:* Monmouth, *Charles.* **3.** Near Caerleon†, *Clark, Hamilton:* Usk, *Hamilton:* Michaelstone y Vedw; near Llanvihangel Llantarnam; Mescoed Mawr, Bettws; Pensylvania, near Castleton; Llandegfedd; Llanddewi Fach; near Plas Machen*; Malpas. **4.** St. Pierre†, *Clark, E. J. Lowe:* between Chepstow and Tintern, *Shoolbred:* Rhyd-y-fedw, *T. G. Evans:* near Milton; Mitchel Troy. **5.** Marshfield*†, *Clark,!:* near Ebbw Bridge; St. Brides Wentlloog, *Hamilton:* Newport, *McKenzie:* Caldicot Level, *Mrs. Ellis:* Percoed Reen, Coedkernew; Whitson; near Llanwern; near Ffynnon Slwt, St. Mellons*.

Lythrum portula (L.) D. A. Webb Water Purslane
Peplis portula L.

Margin of ponds and moorland pools; rare. Native.
1. Llanfoist*. **2.** Mynydd Machen; below Mynydd Dimlaith. **3.** Near Usk Castle, *Clark:* near Castleton Waterworks, *Hamilton:* Wentwood Reservoir. **4.** Fair Oak Pond, near Tintern, *Shoolbred.*

ssp. **longidentata** (Gay) Sell
4. Trelleck Bog*.

THYMELAEACEAE
DAPHNE L.
Daphne mezerion L. Mezerion

Formerly grew on both sides of the Wye near Chepstow, but is now extinct.

114

THYMELAEACEAE

Daphne laureola L.   Spurge Laurel

Woods on calcareous soils; locally frequent. Native.
1. Hendre Wood, *Hamilton:* Lilyrock Wood*; Wonastow Wood*; Lady Park Wood*; *Charles:* Onen*, *Freer.* **2.** Craig-yr-allt Wood, above Pontypool, *Clark.* **4.** Near Rogiet; Mount Ballan Wood, *Hamilton:* Piercefield Woods*; woods by the Wye, *Shoolbred,!.* Wynd Cliff*, *Shoolbred, McKenzie,!:* between Portskewett and Crick*, *Willan:* St. Brides Nertherwent*, *Mrs. Ellis:* Troy Park Wood*, *Charles:* Black Cliff Wood; The Minnetts; near Carrow Hill.

ONAGRACEAE

EPILOBIUM L.

Epilobium hirsutum L.   Great Hairy Willow-herb

Streamsides, ditches and marshes; common. All districts. Native.

Epilobium parviflorum Schreb.   Small-flowered Willow-herb

Marshy places and streamsides; frequent and widespread. All districts. Native.

Epilobium montanum L.   Broad-leaved Willow-herb

Hedgerows, woods, shady streamsides, cultivated ground and waste places; common. All districts. Native.

×**obscurum** (*E.* ×*aggregatum* Celak.)
4. Tintern*, *Ley, Shoolbred.*

×**parviflorum** (*E.* ×*limosum* Schur.)
4. Portskewett*, *Shoolbred.*

Epllobium lanceolatum Seb. & Mauri   Spear-leaved Willow-herb

Roadside banks and walls; rare. Native.
2. Darran Road and Pen-y-rhiw, Risca*; Coed Trinant, Llanhilleth*. **3.** Caerleon*; Rogerstone. **4.** Near Tintern, *Thwaites:* Portskewett Station*, *Marshall:* Magor Station. **5.** Marshfield*.

×**montanum** (*E.* ×*neogradense* Borbás)
1. Dixton*, *Riddelsdell.*

Epilobium roseum Schreb.   Pale Willow-herb

Roadsides, wall bases and cultivated ground; rare. Native.
1. By the River Trothy, Tal-y-coed*; Lady Park Wood*, *Charles:* between Priory Grove and Lady Park Wood; Staunton Road, near Monmouth*, *R. Lewis.* **3.** St. Julians, *Hamilton:* Pont-hir, near Llanfrechfa*. **4.** Tintern*, *Ley, Druce, Roper, Shoolbred:* Shirenewton;

ONAGRACEAE

Mathern; between Pwllmeyric and Mathern*; Parkwall Hill, St. Pierre*, *Shoolbred:* Chepstow, *Shoolbred, Redgrove:* by Trelleck Bog; Lone Lane, Pentwyn, Penallt*, *Charles.* **5.** Nash; St. Brides Wentlloog, *Hamilton.*

Epilobium adenocaulon Haussk.   Glandular-stemmed Willow-herb

Roadsides, cultivated ground, woodland paths and streamsides; common. All districts. Denizen.

A native of North America. First recorded by Noel Sandwith in 1946.

Epilobium tetragonum L.   Square-stalked Willow-herb

Roadsides, quarries, cultivated ground and woodland paths. Native.

ssp. **tetragonum**
Common. All districts.

ssp. **lamyi** F. W. Schultz   Glaucous Willow-herb
Rare.
1. Near Monmouth, *Ley.* **4.** Chepstow*, *Ley, Redgrove.*

Epilobium obscurum Schreb.   Dull-leaved Willow-herb

Marshes, ditches, streamsides, wet moorlands, wet roadsides and woods, usually on acid soils; common in upland districts, frequent elsewhere. All districts. Native.

× **parviflorum** (*E.* × *dacicum* Borbás)
4. The Glyn, Itton*; Bigsweir*, *Shoolbred.*

Epilobium palustre L.   Marsh Willow-herb

Bogs and acid marshes; frequent to common. All districts. Native.

Epilobium nerterioides Cunn.

A native of New Zealand naturalized on damp walls and rocks, and on old colliery tips; rare.
2. Cwm Ffrwd-Oer, Pontypool, *Vernall:* Crosskeys*. **3.** Newport. **4.** By Trelleck Bog, *T. G. Evans.*

First recorded by H. J. Vernall in 1952.

Epilobium angustifolium L.   Rosebay Willow-herb

Open woods, railway banks, quarries, walls and roadsides; common. All districts. Denizen.

First recorded by Edwin Lees in 1851 and rare until the end of the last century.

ONAGRACEAE

OENOTHERA L.

Oenothera biennis L.   Evening Primrose

Waste ground and railway banks; rare. Alien.
2. Between Cwmfelin-fach and Pont Lawrence, *McKenzie*. **4.** Wood below Wynd Cliff, *Shoolbred:* Tintern, *Trapnell*. **5.** Between St. Brides Wentlloog and Marshfield, *Hamilton:* Marshfield, *Richards:* Portskewett, *T. G. Evans.*

Some of the above records may refer to *Oenothera parviflora* which has been confused with *O. biennis*.

Oenothera parviflora L. sensu lato   Small-flowered Evening Primrose

Waste places and railway banks; rare. Alien
1. Wonastow Road, Monmouth*, *Charles*. **2.** Abercarn*. **4.** Whitebrook*; Redbrook, *Charles*.

Oenothera erythrosepala Borbás
  O. lamarkiana auct.

Waste places and roadsides; rare. Alien.
1. Abergavenny*. **2.** Ynysddu*. **4.** Chepstow, *T. G. Evans:* near Mitchel Troy.

CIRCAEA L.

Circaea lutetiana L.   Enchanter's Nightshade

Woods and hedgerows; common. Native. Districts 1 to 4.

Circaea ×intermedia Ehrh.   Intermediate Enchanter's Nightshade
Rocky limestone woods; rare. Native.
2. Penygarn Wood, Pontypool*, *T. H. Thomas:* Craig yr Hafod, near Blaenavon*, *Guile,!*.

Purton's record for *Circaea alpina* from woods in the neighbourhood of Abergavenny (*Midland Flora*, **1**, 54, (1817) doubtless refers to *Circaea intermedia* at Craig yr Hafod.

HALORAGACEAE

MYRIOPHYLLUM L.

Myriophyllum verticillatum L.   Whorled Water-Milfoil

Reens; very rare. Native.
5. Whitewall Common, Magor*, *R. Lewis.*

Included in a list of plants in White's *Guide to Abergavenny*, 1886 but no locality is given, the record is therefore treated as dubious.

HALORAGACEAE

Myriophyllum spicatum L.   Spiked Water-Milfoil
Ditches, reens, rivers and canals; common. Native.
Districts 1, 3, 4 and 5.

GUNNERA L.

Gunnera manicata Linden ex André
Denizen.
3. Banks of the River Usk, near Llangybi*, 1938, *A. Rowland.*

HIPPURIDACEAE
HIPPURIS L.

Hippuris vulgaris L.   Marestail
Ditches and reens; rare. Native.
3. Caerleon, *Hamilton.* **5.** Ditches by the Severn, between Rogiet and
Magor*, *Shoolbred:* Dyffryn, *Hamilton:* Whitewall Common, Magor,
R. Lewis,!: near Collister Pill, Undy*.

CALLITRICHACEAE
CALLITRICHE L.

Callitriche stagnalis Scop.   Water Starwort
Ditches, reens, streams and ponds; common. All districts. Native.

Callitriche platycarpa Ktzg.
Ditches, reens and ponds. Native.
2. Glebe Farm, Bedwas*. **5.** Marshfield*, *Conway:* Mounton*, *Shoolbred:*
near New House, Rumney*.
Probably commoner than the above records suggest.

Callitriche obtusangula Le Gall.
Reens; locally frequent. Native.
4. Mitchel Troy. **5.** Caldicot, *Mrs. Welch:* near New House, Rumney*;
Peterstone Wentlloog*; Marshfield*; Undy*; Magor*.

Callitriche intermedia G. F. Hoffm. *sens. lat.*
Ditches, reens and streams; locally frequent. Native.
1. Honddu and Grwyne Fawr Valleys, *Ley.* **2.** Near Tredegar. **4.** Cross-
way Green Farm, Chepstow; between Portskewett and St. Pierre,
Shoolbred. **5.** Ditches by the Severn, between Magor and Mathern; near
Rogiet, *Shoolbred.*

ssp. **pedunculata** (DC.) Syme
5. Undy*.

ssp. **hamulata** (Kutz.) Clapham
5. Marshfield*; Undy*.

CALLITRICHACEAE

LORANTHACEAE
VISCUM L.

Viscum album L.   Mistletoe

Parasitic on various trees; locally frequent. Native. **1.** Llangattock Lingoed on oak, *Bull, Webster:* near The Hendre, on hazel, *Coomber* ex *Hamilton:* Wyesham and several places about Monmouth, *Charles:* Daren above Cwmyoy, on hawthorn, *Guile.* **3.** Near Usk, on oak; Penpelleni, Goytre, on oak, *Jesse.* **4.** Tintern, *Motley:* near Chepstow Castle, on whitebeam, *E. Lees:* Shirenewton, on ash, apple and hawthorn; Itton Court, on elm, *Lowe:* Dingestow, *Hamilton:* Castle Woods, Chepstow, on lime; Itton; Piercefield, on elm and whitebeam; Golden Hill, St. Arvans, on field maple; Mathern, on hawthorn*, *Shoolbred:* Gwernesney, on white willow, *Durham:* Penterry Farm, Tintern, on apple*, *Charles:* Llandevaud, on Black Italian poplar, *Hyde:* Trelleck Grange, on hazel and hawthorn; Mitchel Troy, on *Robinia*, *Rees:* Llanvair Discoed, on apple*. **5.** Nash, *Hamilton.*

CORNACEAE
SWIDA Opiz

Swida sanguinea (L.) Opiz   Dogwood
   Cornus sanguinea L.

Hedgerows and woods; common. All districts. Native.

ARALIACEAE
HEDERA L.

Hedera helix L.   Ivy

Woods, hedgerows, walls and shaded rocks; common in all districts. Native.

UMBELLIFERAE
HYDROCOTYLE L.

Hydrocotyle vulgaris L.   Marsh Pennywort

Marshes and bogs; common in all districts. Native.

119

UMBELLIFERAE

SANICULA L.

Sanicula europaea L.   Wood Sanicle

Woods and shady hedgerows; common in Districts 1-4. Native.

ASTRANTIA L.

Astrantia major L.

Garden escape.

3. Roadside, Pontnewynydd, 1952*, *Freer*.

ERYNGIUM L.

Eryngium campestre L.

Casual.

4. Roadside near a manure heap, between Tintern Abbey and Tintern Parva, 1853, *Gissing*.

CHAEROPHYLLUM L.

Chaerophyllum temulum L.   Rough Chervil

Hedgebanks and wood border; common. Districts 1, 3-5. Native.

ANTHRISCUS Pers.

Anthriscus caucalis Bieb.   Burr Chervil
  Anthriscus vulgaris Pers.

Rare. Colonist.

4. Wye bank, Bigsweir, *Shoolbred*. **5.** Black Rock, Portskewett, *Shoolbred*.

Anthriscus sylvestris (L.) Hoffm.   Wild Chervil

Hedgebanks and wood borders; very common. All districts. Native.

SCANDIX L.

Scandix pecten-veneris L.   Shepherd's Needle

Cultivated fields; rare. Colonist

1. Kymin Hill, Monmouth*, *Charles*. **2.** Machen, *Hamilton*. **4.** Portskewett, *Hamilton:* Mounton, *Shoolbred:* Windmill Lane, Rogiet, *Mrs. Ellis*. **5.** Dyffryn, *Hamilton*.

Clark in his *Flora of Monmouthshire*, 1868 recorded this plant as abundant in cornfields. Like many other weeds of cultivated land, common in the last century, it has become rare.

UMBELLIFERAE

MYRRHIS Mill.

Myrrhis odorata (L.) Scop.   Sweet Cicely

Roadsides, pastures and waste ground; rare. Denizen.
1. Abergavenny district, *White:* Llanthony district, *Hamilton, Ley:* Grwyne Fawr Valley*, *Ley, Richards:* Pont-y-spig*. **2.** Cwm Ffrwd-Oer, Pontypool, *E. Lees:* near Varteg†; near Trevethin Church; near Pontypool†, *Clark:* Bedwellty Churchyard; Rhyswg Fawr, Abercarn, *McKenzie.* **3.** Pontnewynydd, *Conway.*

TORILIS Adans.

Torilis japonica (Houtt.) DC.   Upright Hedge Parsley
Caucalis anthriscus (L.) Huds.

Hedgebanks and wood borders; common in all districts. Native.

Torilis arvensis (Huds.) Link   Field Parsley
Caucalis arvensis Huds.

Cultivated ground. Colonist.
3. In a cornfield near Raglan, *anon.* 1807.
Hamilton records this as common. If this was so when he published his flora of the county in 1909 it has completely disappeared as a weed of cultivation within the past 50 years.

Torilis nodosa (L.) Gaertn.   Knotted Hedge Parsley
Caucalis nodosa (L.) Scop.

Stony roadsides and sea walls; locally frequent in District 5, rare elsewhere. Native.
3. Caerleon; Bulmore, *Hamilton.* **4.** Llanwern, *Hamilton:* near Penterry, *Shoolbred:* Pwl-pan, near Llanwern. **5.** Liswerry, *Hamilton:* Caldicot and Rogiet Moors*, *Shoolbred:* near Newport Lighthouse, *McKenzie:* near Peterstone, Wentlloog; Rumney*.

CAUCALIS L.

Caucalis latifolia L.

Casual.
1. Cultivated ground, Kymin Hill, Monmouth*, *Charles.*

CORIANDRUM L.

Coriandrum sativum L.   Coriander

Casual.
1. Waste ground, Monmouth*, *Charles.*

121

UMBELLIFERAE

SMYRNIUM L.

Smyrnium olusatrum L. Alexanders

Roadsides, railway and river banks; rare. Denizen.
2. Machen. **3.** Usk, *Miss Frederick:* Bassaleg. **4.** Chepstow, *E. Lees:* Chepstow Castle* and banks of the Wye, Chepstow, *Shoolbred:* Minnetts Lane, *Mrs. Ellis:* Rogiet, *T. G. Evans.*

CONIUM L.

Conium maculatum L. Hemlock

Roadsides, field borders and river banks; frequent. Native.
1. Castle Meadows, Abergavenny*, *Bailey:* near Monmouth, *Whitwell:* Monmouth; Lady Park Wood, *Charles:* Llanthony, *Miss Norman:* Llanfoist; near Great Goytre, Grosmont; Abergavenny*. **2.** Machen; Pontypool, *Hamilton.* **3.** Usk, *Clark:* near St. Julians Wood, Christchurch. **4.** Chepstow; Mounton; Shirenewton; between Chepstow and Shirenewton; near Rogiet*; Bigsweir; Caerwent, *Shoolbred:* Piercefield, *T. G. Evans.* **5.** Between Rogiet and Magor; Caldicot, *Shoolbred:* near Peterstone Wentlloog*; between Marshfield and St. Brides; near Marshfield*.

BUPLEURUM L.

Bupleurum tenuissimum L. Slender Hare's-ear

Salt marshes; very rare. Native.
5. Magor Pill*, *Shoolbred:* Rumney*.

APIUM L.

Apium graveolens L. Wild Celery

Salt marshes and banks of tidal rivers; locally frequent. Native.
3. Near the canal, Allt-yr-yn, *Hamilton.* **4.** River Wye at and above Chepstow, *Clark, Shoolbred:* Mounton; Itton; river bank, Liveoaks*; Tintern, *Shoolbred:* **5.** Between Sudbrook and Portskewett, *Shoolbred:* between Pont Ebbw and the lighthouse, *McKenzie:* near Marshfield*; Magor Pill*.

Apium nodiflorum (L.) Lag. Marshwort

Reens, ditches, pond margins and streamsides; common in all districts. Native.

Apium inundatum (L.) Reichb. f. Lesser Marshwort

Margins of pools; very rare. Native.
1. Llanfoist*. **4.** Trelleck Bog, *Shoolbred.* **5.** Marshfield; St. Brides Wentlloog, *Hamilton.*

UMBELLIFERAE

PETROSELINUM Hill

Petroselinum crispum (Mill.) A. W. Hill   Parsley
  Carum petroselinum (L.) Benth.

Hedgebanks and walls; rare. Alien.
1. Rockfield*, *Charles.* **4.** Pwllmeyric*, *Shoolbred:* Chepstow, *B.S.B.I. Excursion* 1951.

Petroselinum segetum (L.) Koch   Corn Caraway
  Carum segetum (L.) Benth. & Hook.

Roadside banks and pastures; rare. Native. **2.** Machen, *Hamilton.* **3.** Malpas, *Hamilton.* **4.** Near Portskewett railway station*, *Shoolbred:* Bishton*. **5.** Dyffryn, *Hamilton:* near Llanwern; near Rumney*; Goldcliff Pill*.

SISON L.

Sison amomum L.   Stone Parsley

Hedgebanks, roadsides and pastures; locally frequent to locally common especially on the marine alluvium of District 5. Native.
1. Monmouth*, *Riddelsdell, Charles:* between Llanvapley and Abergavenny*; near the Garth; Kymin Hill*; Monmouth; Cross Ash; Rockfield*; Buckholt Wood; near Troy Tunnel*; Leasbrook Lane; Osbaston*; Llantilio Crossenny*; between Onen and Penrhos*; St. Maughan's Green*; Redbrook Road, near Wyesham Halt*, *Charles:* between Abergavenny and Govilon, *Hardaker:* near Pen-isa'r-plwyf Wood, Pandy. **2.** Pontypool, *T. H. Thomas.* **3.** Liswerry, *Hamilton:* Cefntilla, near Raglan*, *Charles:* near Cilfeigan Park; Raglan; Christchurch. **4.** Llandogo, *Watkins:* near Kemeys Commander, *Clark:* Llanwern; Caerwent, *Hamilton:* Langstone; Bishton; Mathern*, *Shoolbred:* Mitchel Troy*, *Charles,* **5.** Magor, *Shoolbred,!:* Caldicot, *T. G. Evans:* between Marshfield and Rumney; about Tredegar Park and St. Brides Wentlloog.

(Purton's record (*Midland Flora,* (1817), **2,** 748) for *Cicuta virosa* L. from ditches about Clytha, and on the Usk near Llanvair is presumably an error for *Oenanthe crocata.*)

CARUM L.

Carum carvi L.   Caraway

Casual.
2. Near Crumlin†, *Clark.*

CONOPODIUM Koch.

Conopodium majus (Gouan) Loret & Barr.   Pignut

Pastures, meadows and woods; common in all districts. Natives.

UMBELLIFERAE

PIMPINELLA L.

Pimpinella saxifraga L. Burnet Saxifrage

Pastures, meadows and roadsides; common. Districts 1-4. Native.

AEGOPODIUM L.

Aegopodium podagraria L. Goutweed

Roadsides and hedgebanks, especially about villages and farms; common in all districts. Denizen.

BERULA Koch

Berula erecta (Huds.) Coville. Narrow-leaved Water Parsnip

Reens, ditches, streamsides and canals; frequent in the south of the county, rare elsewhere. Native.
1. Abergavenny. **3.** Canal side, Newport to Malpas; between Rumney and Llanrumney. **4.** Mounton*; near Mathern Mill; St. Pierre, *Shoolbred:* near Moyne's Court, *H. A. Evans.* **5.** Newport, *Clark:* Tredegar Park, *Hamilton:* Rogiet Moors; Undy, *Shoolbred:* Rumney to Marshfield; Percoed Reen, Coedkernew*; Heol Las, Peterstone Wentlloog*; Rumney*; Goldcliff; near Whitson; Blackwall Reen, Magor*.

(**Sium latifolium** L. recorded by Clark from reen near Marshfield is undoubtedly an error for *Berula erecta.*)

CRITHMUM L.

Crithmum maritimum L. Samphire

Rocks on the coast; rare. Native.
5. Sudbrook*, *Shoolbred, Hamilton, T. G. Evans:* Denny Island, *Matthews.*

OENANTHE L.

Oenanthe fistulosa L. Tubular Water Dropwort

Reens, riversides, canals, ponds and marshes; locally frequent. Native.
3. Maerdy, near Usk; Pontsampit, *Clark:* Caerleon, *Hamilton:* Malpas; Llangattock; Llanddewi Fach*; near Castleton. **4.** Llanllywel, *Clark:* Trelleck, *Ley:* Itton; by the Wye below Tintern, *Shoolbred.* **5.** Near Nash, *Conway:* Marshfield; St. Brides Wentlloog, *Hamilton,!:* between Chapel Tump and Undy*, *Shoolbred:* below Mathern, *Shoolbred, T. G. Evans,!:* Magor, *Shoolbred, T. G. Evans,!:* Llanwern, *T. G. Evans:* Whitson; Rumney to Marshfield*.

(**Oenanthe silaifolia** Bieb. Recorded by A. Ley in the *Botanical Record Club Report.* 1881-82, 191, from damp meadows by the Wye below Tintern. Shoolbred's specimen from the same locality is, however, incorrectly named and is a form of *Oenanthe lachenalii;* Ley's record must therefore be considered a dubious one.)

UMBELLIFERAE

Oenanthe lachenalii C.C. Gmel.   Parsley Water Dropwort
Banks of tidal rivers; rare. Native.
5. Near Nash*, *Conway:* Rumney*.

Oenanthe crocata L.   Hemlock Water Dropwort
Marshes, streamsides and ditches; common in all districts. Native.

Oenanthe aquatica (L.) Poir.   Fine-leaved Water Dropwort
Reens and pond margins; rare to locally frequent. Native.
1. Llanfoist*. **5.** Coedkernew; Marshfield; St. Brides Wentlloog, *Hamilton:* near Magor*, *Nelmes, T. G. Evans:* Undy; near Pîl-du, St. Mellons*; Peterstone Wentlloog*.

AETHUSA L.

Aethusa cynapium L.   Fool's Parsley
Cultivated and waste ground; common in all districts. Native.

FOENICULUM Mill.

Foeniculum vulgare Mill.   Fennel
Railway banks, waste ground and rocky shore of the Severn; rare. Denizen.
1. Monmouth*, *Charles,* **3.** Railway sidings near Cardiff Road, Newport, *Hamilton.* **4.** Tintern, *Ley:* Ifton, *Hamilton:* Chepstow, *Shoolbred.* **5.** Newport*,*Whitwell, Clark,!:* Blackrock, Portskewett, *Mrs. Ellis:* Alexandra Dock, Newport, *Hamilton.*

SILAUM Mill.

Silaum silaus (L.) Schinz. & Thell.   Meadow Sulphur-wort
Pastures and roadsides, chiefly on basic soils and marine alluvium; locally frequent. Native.
1. About Llanthony Abbey, *Conway, Ball:* near Skenfrith, *Trapnell:* Dunkard*; Cross Ash; St. Maughan's*; Buckholt Wood; Onen*; Coedanghred Hill*; Llantilio Crossenny*; between Cross Ash and Skenfrith*; Monmouth; Nant-y-gern, near Newcastle*; Osbaston*, *Charles:* between Abergavenny and Llantilio Pertholey. **3.** near Raglan*, *Charles:* Pontypool Road*; Maindee, Newport; Llangattock; Christchurch. **4.** Wynd Cliff, *Clark:* Chepstow*, *Hamilton, Shoolbred:* Tintern*, *Hamilton, Shoolbred:* Trelleck*, *Shoolbred:* Mitchell Troy; Cwmcarvan*; near Pen-y-clawdd*, *Charles:* Langstone Quarries. **5.** Newport*, *Conway:* Marshfield; Peterstone Wentlloog, *Richards,!:* Magor, *T. G. Evans:* by Greenland Reen, St. Mellons*.

UMBELLIFERAE

ANGELICA L.

Angelica sylvestris L.   Wild Angelica

Marshes, ditches, streamsides and marshy woods; common in all districts. Native.

PASTINACA L.

Pastinaca sativa L.   Parsnip
Peucedanum sativum (L.) Benth. & Hook.f.

Roadsides, reen banks, river banks, pastures and waste ground; fairly common on the marine alluvium of District 5, rare elsewhere. Native. **1.** Monmouth. **3.** Pen-y-lan, near Castleton; between Llantarnam and Pontnewydd*. **4.** Redbrook*, *Charles:* Bishpool. **5.** Ballast banks, Newport†, *Clark, Whitwell:* Magor, *Hamilton, Shoolbred:* reens by the Severn below Undy and Magor Pill, *Shoolbred:* Caldicot Moors, *T. G. Evans:* Llanwern; Peterstone Wentlloog*; Traston, Newport.

HERACLEUM L.

Heracleum sphondylium L.   Hogweed

Roadsides, hedgebanks, pastures and waste ground; common in all districts. Native.

Heracleum mantegazzianum Somm. & Levier

Local. Denizen.
1. Abundant on the bank of the River Usk, Llanelen, 1967*, *Millichamp.*

DAUCUS L.

Daucus carota L. ssp. **carota**   Carrot

Pastures, roadsides, limestone quarries and railway banks, chiefly on basic soils; common in Districts 1, 3 to 5, restricted to the eastern border of District 2. Native.

(ssp. **gummifer** Hook.f. is recorded by Clark from the banks of the Usk at Newport, but this is probably an error.)

CUCURBITACEAE

BRYONIA L.

Bryonia dioica Jacq.   White Bryony

Hedges; common or frequent in Districts 1 and 4, rare elsewhere. Native. **1.** Common. **2.** Near Lower Llanfoist. **3.** Llanover, *Miss Briggs:* Usk; Raglan; between Little Mill and Pontypool Road. **3.** Caerleon*, *Miss Cooke:* Kemeys Commander; Llanddewi Fach. **4.** Ifton; near Caerwent, *Hamilton:* Rogiet, *Hamilton, Charles:* Mounton*, *Shoolbred:* Llanvair Discoed, *Beckerlegge:* Llanmartin*.

EUPHORBIACEAE

MERCURIALIS L.

Mercurialis perennis L. Dog's Mercury

Woods and hedgebanks; common in all districts. Native.

Mercurialis annua L. Annual Dog's Mercury

Cultivated ground, ballast and waste places; rare. Colonist.
1. May Hill, Monmouth*, *Charles:* Abergavenny, *Hardaker.* **3.** Near Bassaleg*. **4.** Tintern*, *Purchas, Ley, Shoolbred:* Shirenewton*, *Shoolbred.* **5.** Newport Docks*, *Conway.*

EUPHORBIA L.

Euphorbia lathyris L. Caper Spurge

Woods and waste places; rare. Denizen.
4. In a wood and near old mines, Runston*; wood at Great Dinham*, *Hamilton, Shoolbred:* Lone Lane, Pentwyn*, *Charles:* Llangoven*, *Lewis Williams.*

Euphorbia serrulata Thuill. Upright Spurge
    E. stricta L. nom. illegit.

Woods and quarries on limestone; rare. Native.
4. Between Wynd Cliff and Tintern*, *Lightfoot, Shoolbred,!:* Llandogo*, *Hamilton, Shoolbred:* near Bigsweir, *Hamilton, Shoolbred.*

First recorded by John Lightfoot in 1773.
According to C. C. Babington (*Memorials*, 1897, 124) *Euphorbia serrulata* was plentiful from above Tintern down the river for several miles. It has now become rare, partly due to the replanting of much of the woodland with conifers.
    A record by Clark in his *Flora of Monmouthshire* for *Euphorbia platyphyllos* from Cefnila probably refers to this species.

Euphorbia helioscopia L. Sun Spurge

Cultivated ground; common in all districts. Colonist.

Euphorbia peplus L. Petty Spurge

Cultivated ground; very common in all districts. Colonist.

Euphorbia exigua L. Dwarf Spurge

Cultivated ground; frequent to locally common. All districts. Colonist.
1. Common about Monmouth, *Charles.* **2.** Pontypool*, *Thomas:* Machen, *Hamilton.* **3.** Malpas, *Hamilton:* Christchurch*; Pen-y-lan, near Castleton*; Llandegfedd. **4.** Common, *Shoolbred:* Chepstow; Usk Road, near Chepstow, *Shoolbred:* Rogiet*, *Mrs. Ellis!:* near The Minnetts. **5.** Coedkernew, *Hamilton.*

EUPHORBIACEAE

Euphorbia ×**pseudovirgata** (Schur.) Soó
E. esula auct.
Alien.
4. Redbrook*, *c.* 1904, *Ley.*

Euphorbia amygdaloides L.   Wood Spurge
Woods; common on the limestone and Old Red Sandstone of Districts 2, 3 and 4; rare in District 2 whence it is recorded from Abercarn. Absent from District 5. Native.

RICINUS L.
Ricinus communis L.   Castor Oil Plant
Casual.
5. Newport Docks, 1942, *Macqueen.*

POLYGONACEAE
POLYGONUM L.
Polygonum aviculare L.   Knotweed
P. heterophyllum Lindm.
Cultivated ground, roadsides and waste places; very common in all districts. Native.

Polygonum arenastrum Bor.
P. aequale Lindm.
Waste ground, roadsides, paths and cart tracks; common in all districts. Native.

(**Polygonum maritimum** L. and **P. raii** Bab. recorded by Hamilton and Clark respectively must, in the absence of confirmation, be treated as errors.)

Polygonum bistorta L.   Bistort
Roadsides, hedgebanks and damp meadows; widely distributed but not common. Native.
1. Grwyne Fawr Valley; Cwmyoy; near Pont-y-spig, *Ley:* Llanthony*, *Richards:* Dixton*; Trivor, near Skenfrith*; Hadnock Road, Monmouth; between Wyesham and The Kymin, *Charles:* near Hadnock, *Sandwith:* Abergavenny Junction*; near Coed-y-fedw. **2.** Gelligroes Mill, *McKenzie:* near Llwyn-llynfa, Bedwas. **3.** Llanllywel; Llangybi, *Clark:* Pen-y-garn, Pontypool*, *T. H. Thomas:* Tovey's Farm, Henllys, *Hamilton:* Raglan district, *Miss Frederick:* near Bettws; near Henllys Wood; Castell-y-bwch; between Panteg and Pontypool Station; Penpelleni*. **4.** New-church*, *Clark, Shoolbred:* Tintern*; Penterry; The Glyn; Llandogo; Magor*, *Shoolbred:* Upper Redbrook*; Hael Wood, Penallt*, *Charles:* Trelleck, *Young:* The Coombe, Shirenewton, *Mrs. Ellis:* Mounton; Fairoak, *T. G. Evans:* Cleddon*.

Polygonum amphibium L.   Water Bistort

Ponds, rivers, streams and reens, and on roadsides as the terrestrial state; frequent, Native.
1. River Wye, between Monmouth and Hadnock*, near Monmouth Cap*, *Charles:* Llanfoist. **2.** Griffithstown. **3.** Newport*, *Conway:* Olway Brook, near Usk, *Clark:* Raglan district, *Miss Frederick:* Allt-yr-yn*; canal, Malpas; Christchurch. **4.** Newchurch; Bigsweir, *Shoolbred:* Whitebrook*, *Charles.* **5.** St. Brides Wentlloog, *Hamilton:* Marshfield, Rumney; Llanwern; Whitson; Undy.

Polygonum persicaria L.   Common Persicaria

Cultivated ground, waste places, stream and pond margins, and river banks; common in all districts. Native.

Polygonum lapathifolium L.   Pale Persicaria
  P. nodosum Pers.

Cultivated ground, river banks and waste ground; frequent. All districts. Native.

Polygonum hydropiper L.   Water Pepper

Streamsides, pond margins, marshes and wet roadsides; common in all districts. Native.

Polygonum mite Schrank   Loose-flowered Persicaria

River banks; rare. Native.
1. Near Wye Bridge, Monmouth*; *Charles:* Hadnock*, *R. Lewis.* **4.** Bigsweir, *Ley:* Llandogo*, *Shoolbred:* between Whitebrook and Penallt*, *Charles.*

  ×**persicaria** (*P.* ×*condensatum* (F. W. Schultz) F. W. Schultz)

1. Bank of the River Wye, Hadnock*; bank of the River Monnow, near Monmouth Cap*, *Charles.* **4.** Between Whitebrook and Penallt, *Charles.*

Polygonum convolvulus L.   Black Bindweed

Cultivated ground and waste places; common. Districts 1, 3-5. Native.

var. **subalatum** Lej. & Court.
1. Fiddler's Elbow, near Monmouth, *R. Lewis.* **4.** Common in the Chepstow district; Pandy Mill, Itton*, *Shoolbred.* **5.** Between Castleton and Marshfield*.

(**Polygonum dumetorum** L. Recorded for Monmouthshire in Watson, *Topographical Botany*, ed. 1, 1873, on the authority of W. H. Purchas. The record is unconfirmed and may be an error for *P. convolvulus* var. *subalatum*.)

POLYGONACEAE

Polygonum cuspidatum Sieb. & Zucc.   Japanese Fleece-flower

Common in many localities on roadsides, railway and river banks in all districts. Denizen.

Polygonum polystachyum Wall. ex Meisn.

Garden escape.

4. Bulwark, Chepstow, 1942, *Mrs. Ellis.*

FAGOPYRUM Mill.

Fagopyrum esculentum Moench   Buckwheat

Waste ground and open woods; rare. Casual.

1. Kymin Hill, Monmouth*, *Charles.* **4.** Pen-y-clawdd*, *Miss C. Parry:* Raven's Nest Wood, Tintern*, *Shoolbred.*

Fagopyrum tataricum (L.) Gaertn.

Casual.

4. Raven's Nest Wood, Tintern*, *Shoolbred.*

RUMEX L.

Rumex acetosella L. sens. lat.   Sheep's Sorrel

Dry pastures and banks, heaths, waste ground and walls; common in all districts. Native.

Rumex acetosa L.   Sorrel

Meadows, pastures and grassy roadsides; common in all districts. Native.

Rumex scutatus L.

Casual.

4. Tintern, *Woods.*

Rumex hydrolapathum Huds.   Great Water Dock

Reens; locally frequent. Native.

4. Between Mathern and St. Pierre, *Shoolbred.* **5.** Wentlloog reens, *Storrie:* near the lighthouse, Newport, *McKenzie:* Peterstone Wentlloog*; Percoed Reen, Coedkernew*; Llanwern; Magor.

Rumex cristatus DC.

R. graecus Boiss. & Heldr.

Denizen; rare.

3. Bank of the River Rhymney, Rumney, where it is known to have occurred on both sides of the river for nearly 50 years.

Rumex crispus L.   Curled Dock

Roadsides, waste ground, cultivated ground and tidal banks of rivers; very common in all districts. Native.

× **obtusifolius** (*R.* × *acutus* L.)
4. Chepstow; Portskewett, *Shoolbred.* **5.** Near Marshfield*.

Rumex obtusifolius L.   Broad-leaved Dock

Roadsides, hedgerows, meadows and waste ground; very common in all districts. Native.

Rumex sanguineus L.
var. **sanguineus**
Denizen.
4. Near Cottages, Mounton*, *Shoolbred.*

var. **viridis** Sibth.   Green-veined Dock
Hedgerows and woods; common in all districts. Native.

Rumex pulcher L.   Fiddle Dock
Roadsides; rare. Denizen.
4. Castle Dell, Chepstow; Runston*, *Shoolbred.*

Rumex conglomeratus Murr.   Sharp Dock
Banks of rivers and ponds, and marshes; common in all districts. Native.

× **pulcher** (*R.* × *muretii* Hausskn.)
4. Castle Dell, Chepstow*, *Shoolbred.*

Rumex palustris Sm.   Marsh Dock
Reensides and marshes; rare. Native.
5. Blackwall Reen, Magor*, *E. Nelmes,!:* Undy*.

URTICACEAE
PARIETARIA L.
Parietaria judaica L.   Pellitory-of-the-wall
   P. officinalis auct. non L., *P. ramiflora* Moench.

Walls and rocks; common, especially on the limestone. **All districts.** Native.

URTICACEAE

URTICA L.

Urtica urens L. Small Nettle

Cultivated ground, roadsides near houses, and farmyards; frequent. Native.
1. Near Abergavenny. **2.** Ebbw Vale*. **4.** Caerwent*, *Downing:* Thornwell, Chepstow*; St. Arvans Grange; Rogiet; near Severn Tunnel Junction, *Shoolbred:* Trelleck, *Charles:* Tintern, *Webb.* **5.** Between Marshfield and Castleton.

Urtica dioica L. Nettle

Roadsides, hedgerows, wood borders and waste places; very common in all districts. Native.

(Urtica pilulifera L. An unlocalized specimen in Herb. C. T. & E. Vachell is believed to have been collected at Caerwent.)

CANNABIACEAE

HUMULUS L.

Humulus lupulus L. Hop

Hedges and thickets; common. All districts. Native.

ULMACEAE

ULMUS L.

Ulmus glabra Huds. Wych Elm
U. scabra Mill.; *U. montana* Stokes

Woods and hedgerows; common. Districts 1-4. Native.
× **plotii** (*U.* × *elegantissima* Horwood)
4. Near quarry beyond Well House, near Chepstow*, *Shoolbred.*

Ulmus procera Salisb. English Elm
U. campestris auct.

Hedgerows; common in Districts 1, 3 and 4, frequent in Districts 2 and 5. Denizen, often planted.

Ulmus × **hollandica** (Mill.))Moss Dutch Elm
U. coritana × *glabra*

Woods and hedgerows; rare. Denizen.
1. 1¼ miles from Monmouth on the Rockfield Road; pasture south of Llangattock Vibon Avel Manor, *Ley:* Abergavenny, *Druce:* Tintern, *Webb.*
4. Trelleck, *Ley:* between Bishton and Llanwern*.

ULMACEAE

Ulmus plotii Druce   Plot's Elm

Probably planted.

3. One tree in a field at Llanfair Cilgedin*, *Shaw, Nelmes.*

JUGLANDACEAE
JUGLANS L.

Juglans regia L.   Walnut

Woods; rare. Denizen.

3. Bettws Newydd*. **4.** Limestone wood, Castle Wood, Chepstow*, *Shoolbred.*

BETULACEAE
BETULA L.

Betula pendula Roth   Silver Birch
  B. alba auct.

Woods, copses and hedgerows; common in Districts 1-4, unrecorded from District 5. Native.

× **pubescens** (*B.* × *aurata* Borkh.)
2. Cwm Gwyddon, Abercarn*.

Betula pubescens Ehrh.   Downy Birch

Woods, copses and hedgerows; common in the upland areas of Districts 1 and 2, frequent in Districts 3 and 4. Unrecorded from District 5. Native.

var. **glabrata** Wahl.

1. Tarren yr Esgob*. **2.** Cwm Carn, Abercarn*. **4.** Raven's Nest Wood, near Tintern, *Shoolbred:* Llanmelin Camp, *Hyde.*

ALNUS Mill.

Alnus glutinosa (L.) Gaertn.   Alder
  A. rotundifolia Stokes

River and streamsides, wet woods and thickets, and marshes; common in all districts. Native.

CORYLACEAE
CARPINUS L.

Carpinus betulus L.   Hornbeam

Woods; rare. Probably native in the Wye Valley, planted elsewhere.
1. On the Hereford Road, near Abergavenny*, *Bailey:* Llanthony; The Hendre, *Hamilton:* Wyesham*; Harper's Grove, Monmouth; Halfway House Wood, Monmouth*, *Charles:* Garth Wood*, *R. Lewis:* Govilon.

CORYLACEAE

3. Llanover Park, *Hamilton:* near New Inn, near Pontypool Road. **4.** In a remnant of native wood by the railway south of Chepstow Station*; between Pen y Van and Bigsweir; several large trees, probably planted, in Piercefield Park, *Shoolbred,!:* Llanwern; Lawrence Hill, *Hamilton.*

CORYLUS L.

Corylus avellana L.   Hazel

Woods, thickets and hedgerows; common throughout the county. Native.

FAGACEAE

FAGUS L.

Fagus sylvatica L.   Beech

Woods and old hedgerows; common and native on the limestone, common as a planted or self-sown tree elsewhere.

CASTANEA Mill.

Castanea sativa Mill.   Sweet Chestnut

Woods; frequently planted but self-sown trees occur. Denizen.
1. Lady Park Wood, *Charles.* **3.** The Forest, Llangybi. **4.** Piercefield Woods; Hael Woods, Whitebrook; Livox Wood, Penallt.

QUERCUS L.

Quercus cerris L.   Turkey Oak

Woods; rare. Denizen.
4. Wood border, Barnett Woods*; Usk Road, near Chepstow*, *Shoolbred:* railway cutting, Chepstow, *T. G. Evans.*

Quercus ilex L.   Holm Oak

Woods; rare. Denizen.
4. Wooded limestone cliffs, Castle Wood, Chepstow*, *Shoolbred.*

Quercus robur L.   Pedunculate Oak

Woods and hedgerows; common. All districts. Native.

Quercus petraea (Mattuschka) Liebl.   Sessile Oak

Woods; common in the upland areas, frequent elsewhere. Districts 1-4. Native.

×**robur** (*Q.* ×*rosacea* Bechst.)   Hybrid Oak
A noteworthy specimen of the Hybrid Oak is the Caerhyder Oak at Pencreeg Farm, Llanhennock. In 1931 it measured 24 ft. 10 in. in girth at 6 ft. and 45 ft. at ground level, and had 88 ft. span of branches.

SALICACEAE
POPULUS L.

Populus alba L. White Poplar

Hedgerows; frequent as a planted tree.
1. By the River Monnow at The Forge, Monmouth*; Vauxhall Meadows, Monmouth; by the River Monnow, near Rockfield Road, Monmouth*; Barn Farm, Hadnock Road, near Monmouth*; Wyesham*, *Charles*. **3.** Ponthir, *Hamilton:* Raglan district, *Miss Frederick:* Llanddewi Fach. **4.** Near Trelleck; between Brockweir and Bigsweir, *Shoolbred:* Pont y Saison, *T. G. Evans*. **5.** Marshfield, *Hamilton*.

Populus canescens (Ait.) Sm. Grey Poplar

Woods and hedgerows; rare. Usually planted though possibly native.
1. By the River Monnow, Osbaston*; King's Wood, Wonastow*; White House, near Grosmont*; Lydart Road, near Monmouth, *Charles*. **3.** Castleton*. **4.** Between Moynes Court and St. Pierre*, *Hyde*.

Populus tremula L. Aspen

Woods and hedgerows; locally frequent. Native.
1. Llanthony, *Ley:* Wyesham; between Wyesham and The Kymin, *Charles:* Rockfield, *Collett:* Tal-y-coed*, *Hyde:* Seargent's Grove, Monmouth*. **2.** Abercarn; Lasgarn Wood, Abersychan. **3.** Pontyclivon, Usk, *Clark:* Llwyncelyn Woods, near Tredunnock, *Hamilton:* between Malpas and Bettws; Lady Hill Wood, Gwehelog. **4.** Earlswood Common*; Barnett Woods; Chepstow Park, *Shoolbred*.

SALIX L.

Salix alba L. White Willow

Banks of rivers, streams, and reens; frequent in District 5, rare elsewhere. Native.
1. About Monmouth, *Charles*. **3.** Llanddewi Fach*. **4.** Mounton*; Mathern, *Shoolbred:* Upper Redbrook*, *Charles:* Langstone. **5.** Below Rogiet, Undy* and Magor*, *Shoolbred:* Caldicot Moor, *T. G. Evans:* Peterstone Wentlloog to Rumney*; near Llanwern.

var. **vitellina** (L.) Stokes
5. Near Chapel Tump Farm, Undy, *Shoolbred*.

× **fragilis** (*S.* × *rubens* Schrank)
5. Several large old trees, apparently this hybrid, on the moors below Magor, *Shoolbred*.

Salix fragilis L. Crack Willow

River banks and streamsides; common. All districts. Native.

SALICACEAE

var. **decipiens** (Hoffm.) Koch
4. Mounton Valley*, *Marshall and Shoolbred.*

Salix triandra L.   Almond Willow
Banks of rivers, streams and reens; locally frequent. Native.
1. By the River Usk, Abergavenny, *E. Lees,!:* near Pont-y-spig, probably planted, *Ley:* Llangattock Vibon Avel*, River Wye, above Monmouth*, *Charles:* by the River Wye, Dixton Newton Parish, *R. Lewis:* near Llanfoist*. **3.** By the Afon Lwyd, between Pontnewydd and Llanfihangel Llantarnam*. **4.** By the River Wye from Llandogo to Whitebrook*, *Shoolbred, Charles, Webb.* **5.** Caldicot and Undy Moors, *Shoolbred.*

× **viminalis** (*S.* × *mollissima* Ehrh.)
4. Mounton*; between Brockweir and Coed Ithel*, *Shoolbred.*

Salix purpurea L.   Purple Willow
River banks; rare. Native.
1. Near Pont-y-spig, *Ley:* Abergavenny*. **3.** Bassaleg*; by the River Usk, Llanllywel*. **4.** Portskewett, *Shoolbred:* pond near the Bulwark, *Miss M. Cobbe.*

× **viminalis** (*S.* × *rubra* Huds.)
4. Between Brockweir and Bigsweir, *Shoolbred.*

Salix viminalis L.   Osier
River banks, stream and reen sides; frequent in Districts 1, 3-5, rare in District 2. Native.

Salix caprea L.   Sallow, Goat Willow
Damp woods, marshy thickets, hedgerows and banks of streams; common. All districts. Native.

× **cinerea** (*S.* × *riechardtii* A. Kerner)
3. Pwll Diwaelog, Castleton*. **4.** Black Cliff Wood, Tintern.

× **viminalis** (*S.* × *laurina* Sm.)
1. Priory Grove, Dixton Newton*, *R. Lewis.*

Salix cinerea ssp. **oleifolia** Macreight   Grey Sallow
Banks of streams, reens and rivers, wet woods and marshes: common. All districts. Native.

× **viminalis** (*S.* × *smithiana* Willd.)
1. Between Pandy and Oldcastle*; Abergavenny*. **3.** Llanddewi Fach*. **5.** Magor; Undy, *Shoolbred:* near Magor*.

SALICACEAE

Salix aurita L.   Auricled Sallow

Damp woods, copses, heaths and marshy ground, usually on acid soils; locally frequent. Native.
1. Pont-y-spig, *Ley,!.* **2.** Trefil; Cwm Gwyddon, Abercarn*; Cwm Tysswg, Abertysswg*; near Maes-y-cwmmer; Pen-y-lan Pond. **3.** Glascoed, *Clark:* Old Reservoir, near Pontnewydd; near New Inn, Pontypool Road. **4.** Near Chepstow; Penterry*; Yellow Moor, near Tintern; Chepstow Park Wood; Devauden; between Tintern and Trelleck, *Shoolbred:* Trelleck Bog*, *Shoolbred, R. Lewis,!.*

× **cinerea** (*S.* × *multinervis* Doell)
4. Boggy ground below Chepstow Park*, *Shoolbred:* Trelleck Bog*.

Salix repens L.   Dwarf Willow

Heaths and boggy places; rare. Native.
4. Boggy heath below Coed Cae, near Shirenewton*, *Shoolbred:* Trelleck Bog; Pen-y-fan, near Whitebrook*, *Charles.*

ERICACEAE

CALLUNA Salisb.

Calluna vulgaris (L.) Hull   Ling,   Heather

Heaths, moors, open woods and banks on acid soils; common in District 2 and on the higher ground of Districts 1 and 4, rare in District 3. Absent from District 5. Native.
In District 3 it has been recorded from near Little Creigydd, Llanddewi Fach, *Campbell:* Garw Wood, near Pontnewydd; near Pant-yr-eos and near Castell-y-bwch.

var. **pubescens** Hull
4. Whitelye Common*.

ERICA L.

Erica tetralix L.   Cross-leaved Heath

Bogs and wet moorland; locally common. Native.
1. Common in the Honddu and Grwyne Fawr Valleys. **2.** Common. **3.** Garw Wood, near Pontnewydd. **4.** Hills above Tintern; The Fedw; Itton Common; Trelleck Bog*, *Shoolbred,!:* Pen-y-fan, near Whitebrook*, *Charles:* near Lydart, Penallt, *Beckerlegge:* Whitelye Common.

Erica cinerea L.   Bell Heather

Heaths, woods and banks; locally frequent. Native.
1. Sugar Loaf, *E. Lees.* **2.** Blorenge, *Hamilton:* Mynydd Maen, *Miss Briggs.* **4.** Chepstow Park; Kilgwrrwg*; Itton; Earlswood; between

ERICACEAE

Rogiet and St. Brides; Wentwood; near Tintern, *Shoolbred:* Trelleck Bog*; The Narth*, *Charles:* The Minnetts, *Mrs. Ellis:* near Lydart, Penallt, *Beckerlegge:* near Slade Wood*.

VACCINIUM L.

Vaccinium vitis-idaea L.   Cowberry

High moorlands; rare, except in the Honddu Valley where it is locally common. Native.
1. About Llanthony, *Ball, Hamilton:* Honddu and Grwyne Fawr Valleys; Hatterals; Ffwddog, *Ley:* between Bal Mawr and Chwarel-y-Fan, Black Mountains*, *Charles*. **2.** Blorenge, *Conway.!:* Garn-ddyrys, *Conway:* Mynydd-y-garn-fawr, Blaenavon, *Clark, T. H. Thomas,!:* Craig yr Hafod, near Blaenavon, *Guile*.

Vaccinium myrtillus L.   Bilberry

Heaths, open woods and moorlands on acid soils; common. Districts 1-4. Native.

Vaccinium oxycoccus L.   Cranberry
Oxycoccus quadripetala Gilib.; *O. palustris* Pers.

Bogs, very rare. Native.
2. Above Varteg*, *Conway:* Blaenavon, *Clark:* northern part of the Rhymney Valley, *Hamilton*. **4.** Trelleck Bog*; *Ley, Hamilton, Shoolbred, Charles,!*.

PYROLACEAE
PYROLA L.

Pyrola minor L.   Lesser Wintergreen

Woods, rare. Native.
1. Below the Coleford Road between the War Fields and the 'Duke of York'; near Martin's Pool, Monmouth; Beaulieu Wood*, *Charles*. **2.** Blorenge, *Hamilton*. **4.** Wentwood, *Clark:* Llandogo Glen, *Ley:* Wynd Cliff*, *Ley, Shoolbred,!:* Piercefield Woods; between Pen-y-parc and Devauden, *Shoolbred:* The Barnetts, *Worsley-Benison:* Pen-y-cae-mawr, Wentwood*; Chepstow Park Wood, *Rickards:* wood by Trelleck Bog, *Charles:* Penallt, *Mrs. Langford*.

(Records of *Pyrola media* and *Pyrola rotundifolia* are errors.)

PYROLACEAE

ORTHILIA Raf.

Orthilia secunda (L.) House   Toothed Wintergreen
Pyrola secunda L.; *Ramischia secunda* (L.) Garcke

Rocky limestone woods; very rare. Native.
4. Near Wynd Cliff, *Bickham, Ley*.

First recorded by S. H. Bickham in 1845 and later by the Rev. A. Ley in 1876. There are no later records and it is probably now extinct.

MONOTROPACEAE

MONOTROPA L.

Monotropa hypophegea Wallr.   Yellow Bird's nest
Monotropa hypopitys L. var. *glabra* Roth

Woods, very rare. Native.
4. Under beeches, near Black Cliff, near Tintern*, *Shoolbred,!.*

EMPETRACEAE

EMPETRUM L.

Empetrum nigrum L.   Crowberry

Moors and bogs; locally common. Native.
1. Hatteral and Ffwddog ranges of the Black Mountains, *Duncan:* Honddu and Grwyne Fawr Valleys, *Ley:* near Trewyn, *Dr. Wood:* Goder Ridge, Black Mountains, *Richards:* Tarren yr Esgob*, *Charles,!.*
2. Mynydd-y-garn-fawr, Blaenavon*, *Clark,!:* New Tredegar*, *James:* Blorenge*; Mynydd Garn Clochdy; Trefil*; Mynydd Bedwellte; Cefn Manmoel*; Ebbw Vale; Mynydd Carn-y cefn*; Mynydd James; Trefil Ddu*; Mynydd Llwyd, near Pontypool; Buarth Maen, between Pontypool and Crumlin*; Aberystruth*. **4.** Trelleck Bog*, *Mrs. Welch*.

PLUMBAGINACEAE

LIMONIUM Mill.

Limonium vulgare Mill.   Sea Lavender

Salt marshes; rare. Native.
5. Magor, *T. G. Evans*.

ARMERIA Willd.

Armeria maritima (Mill.) Willd.   Thrift

Salt marshes and saltings; common along the Severn. Native.

PRIMULACEAE
PRIMULA L.
Primula veris L.   Cowslip

Meadows, pastures, open woods and hedgebanks; frequent. Districts 1, 3 and 4. Native.

Much less frequent than formerly owing to the ploughing and re-seeding of old pasture.

× **vulgaris**

1. Near the 'Duke of York' Inn, Coleford Road, Monmouth*, *Charles.*
4. Dinham; Shirenewton; railway bank below Mathern*, *Shoolbred:* near Howick, *H. A. Evans:* Ifton Great Wood, *Mrs. Ellis:* The Coombe; Llangwm, *T. G. Evans.*

Records for *P. elatior* by Conway and Hamilton from near Pontnewydd and Monmouth district respectively refer either to this hybrid or to the caulescent form of *P. vulgaris.*

Primula vulgaris Huds.   Primrose

Woods, copses, hedgebanks, railway banks and pastures; common in Districts 1, 3 and 4; locally common in District 2; rare in District 5. Native.

CYCLAMEN L.
Cyclamen hederaefolium Ait.

Alien.

4. Hedgebank, Usk Road, St. Arvans*, 1920, *Shoolbred, Redgrove.*

LYSIMACHIA L.
Lysimachia nemorum L.   Wood Pimpernel

Woods and hedgebanks; common. All districts. Native.

Lysimachia nummularia L.   Creeping Jenny, Moneywort

Damp hedgebanks, wet woods, marshes, river banks and stream sides; locally frequent. Native.
1. Newton Dixton; Monmouth*; Leasebrook Lane, Monmouth*; Hadnock; Osbaston*; Rockfield*; Wonastow*, *Charles:* near Llanfoist; near Little Goytre, near Pandy. **3.** Usk, *Clark:* Pant-yr-eos, *Hamilton:* Raglan district, *Miss Frederick,!:* Llandenny, *Charles:* by the Afon Lwyd, Llanfihangel Llantarnam; Malpas; Plas Machen. **4.** St. Pierre, *Clark:* Portskewett; Llanvair Discoed; near Tintern; Trelleck*; Mounton, *Shoolbred:* Lone Lane, Pentwyn, Penallt*; Trelleck Bog*; near Tintern Abbey; Whitebrook Valley*; Cwmcarvan*, *Charles:* between Newchurch

PRIMULACEAE

and Cross-way Farm, *Lowe:* The Coombe, Shirenewton, *Mrs. Ellis:* between Llanwern and Bishton; Buckle Wood. **5.** Between Llanwern and Whitson; near Quarry Hill, St. Mellons*; Undy.

Lysimachia vulgaris L.   Yellow Loosestrife

Marshes, alder woods, river banks, streamsides and pond margins; locally frequent. Native. **1.** Banks of the River Monnow, near Troy House, *Woods:* alder wood, Pont-y-spig, *Ley,!:* Dixton*; The Gamblins, Monmouth*; near Boy's Rocks, Hadnock*, *Charles:* near Llanfoist. **2.** Bank of the River Rhymney, Lower Machen, *Hamilton.* **3.** Usk, *Clark, Mrs. Scholberg:* Castleton, *Hamilton,!:* Garw Wood, near Pontnewydd; near Pentwyn Farm, Llanddewi Fach*; near Castell-y-bwch; near Plas Machen*. **4.** Rhyd-y-fedw; Shirenewton*, *Shoolbred:* Trelleck, *Perman:* Bigsweir*; Penallt*; near Redbrook*; Whitebrook*, *Charles.* **5.** St. Brides Wentlloog, *Hamilton:* Llanwern, *T. G. Evans:* Whitson*; Marshfield* near Ffynnon Slwt, St. Mellons*.

GLAUX L.

Glaux maritima L.   Sea Milkwort

Salt marshes and mud flats; common along the shore of the Severn and the mouths of the rivers. District 5. Native.

ANAGALLIS L.

Anagallis tenella (L.) L.   Bog Pimpernel

Bogs and acid marshes; locally frequent. Native. **1.** Black Mountains, *Richards:* Sugar Loaf*, *Lewis:* Triley Bridge, near Llantilio Pertholey; near Pont-y-spig; Llanfoist. **2.** Abersychan†, *Clark:* Cwmcarn; between Blackwood and Newbridge, *McKenzie:* Pontllanfraith, *Mrs. Leney:* Mynydd Machen; Mynydd Dimlaith; Cwm Carno, near Brynmawr; Trefil. **3.** Near Foxwood; Ynys-y-fro, *Hamilton:* Michaelstone Bridge*, *Richards:* Castell-y-bwch, Henllys, *Sculthorpe,!:* Garw Wood, near Pontnewydd; near Pentwyn Farm, Llanddewi Fach*. **4.** Earlswood†; Pen-y-cae-mawr, Wentwood, *Clark:* between Monmouth and Trelleck, *Ley:* near Tintern; Chepstow Park Wood; Coed Cae, Shirenewton*, *Shoolbred:* Trelleck Bog, *Shoolbred,!:* Pen-y-fan, near Whitebrook*, *Charles:* Penallt Common, *Beckerlegge,!:* Llwynycelyn, *T. G. Evans.*

Anagallis arvensis L.   Pimpernell

Cultivated ground; common. All districts. Native.

The blue-flowered form, forma *azurea* Hyll. is recorded from several localities around Monmouth, from Penyclawdd (*Miss Davis*) and from Trelleck (*Mrs. Pickington, Charles*).

PRIMULACEAE

SAMOLUS L.

Samolus valerandi L. Brookweed

Stream and reen sides; rare. Native.
1. Bunjup's Brook, *Charles*. **4.** St. Pierre, *Clark, Shoolbred:* near Mathern,*
Shoolbred. **5.** Marshfield, *Hamilton:* Peterstone Wentlloog*.

BUDDLEJACEAE

BUDDLEJA L.

Buddleja davidii Franch. Buddleia

Roadsides, railway banks and walls; rare. Established garden escape.
2. Risca. **3.** Usk, *Miss E. A. Jenkins.* **4.** Undy, 1945; Chepstow. **5.** Spencer
Steel Works site; Rumney.

OLEACEAE

FRAXINUS L.

Fraxinus excelsior L. Ash

Woods and hedgerows; common. All districts. Native.

SYRINGA L.

Syringa vulgaris L. Lilac

Self sown or a relic of cultivation in hedgerows.
4. Near Milton. **5.** Near St. Brides Wentlloog.

LIGUSTRUM L.

Ligustrum vulgare L. Privet

Hedgerows, thickets and wood borders; common on the limestone and
marine alluvium, frequent or rare elsewhere. Native.
Distribution uncertain since the commonly planted *Syringa ovalifolium*
has been mistaken for it.

APOCYNACEAE

VINCA L.

Vinca minor L. Lesser Periwinkle

Hedgebanks and wood borders; rare. Denizen.
1. Llanthony district, *Hamilton:* Roadside near Garth; Rockfield*;
Osbaston*; Pont y Saison*, *Charles:* Skenfrith, *Miss Frederick.* **2.** Wood
Miller's Road, Pontypool, *T. H. Thomas.* **3.** Llanllywel, *Clark.* **4.** Went-
wood, *Clark:* Llanwern, *Hamilton:* Tintern, *Monington, Shoolbred:*
Whitebrook*, *Shoolbred:* Coal Pit Farm, Shirenewton, *Lowe:* Hael Wood,
Penallt*, *Charles:* between Tintern and Trelleck Grange*.

142

APOCYNACEAE

Vinca major L.　Greater Periwinkle

Hedgebanks, an escape from cultivation; locally frequent.
1. Abergavenny district, *White:* Dixton*; Wyesham; near Skenfrith*;
Buckholt*; Norton*; near Hadnock*; Wonastow*; near the Onen*;
Ancre Hill, Rockfield; Osbaston*; Brynderi, near Llantilio Crossenny,
Charles: between Skenfrith and Abergavenny, *Lousley:* Tregare. **3.** Usk,
Clark: Llanover, *Briggs:* Henllys Vale; Malpas Road, Newport; near
Coed y Bedw; between St. Mellons and Michaelstone y Vedw*; Castleton*.
4. Chepstow; Caldicot, *Clark:* Tintern, *Shoolbred:* Pen-y-clawdd*; Lone
Lane, Pentwyn*; Coed-y-fedw, near Raglan; Trelleck, *Charles:* near
Milton*; near Penallt Old Church. **5.** Between Marshfield and Castleton;
Goldcliff; near Pye Corner.

(Hamilton's records for *Vinca minor* from near Stud Farm, Coedkernew and
Castleton are probable errors for *V. major*.)

GENTIANACEAE

CENTAURIUM Hill

Centaurium pulchellum (Sw.) Druce

Rare. Colonist.
4. Cultivated field, Llanolway*, *Miss Harrhy.*

Centaurium erythraea Rafn.　Centaury
　C. umbellatum auct.

Roadsides, pastures, quarries, railway tracks and open woods; fairly
common on the limestone, occasional on the Old Red Sandstone. All
districts. Native.

BLACKSTONIA Huds.

Blackstonia perfoliata (L.) Huds.　Yellow-wort

Pastures, roadsides, dry banks, quarries, railway banks and open woods
on the limestone; locally common in Districts 1 and 4; rare in Districts
2 and 3. Native.
1. Common on the limestone. **2.** Machen; Risca. **3.** Glen Usk Demesne;
near Cwrt Bleddyn, Llangybi, *Clark:* Llanfrechfa, *Hamilton:* near Cilfeigan Park*. **4.** Common on the limestone.

GENTIANELLA Moench

Gentianella campestris (L.) Börner　Field Gentian
　Gentiana campestris L.

Pastures; rare. Native.
1. Cwm Bychel, near Llanthony, *Miss Gee.* **4.** Catbrook*, *Ley, Charles:*
Severn Tunnel Junction; Rogiet; Portskewett, *Hamilton:* near Broad-
stone, between Tintern and Trelleck*, *Shoolbred.*

143

GENTIANACEAE

Gentianella amarella (L.) Börner sens. lat.   Autumn Gentian
  Gentiana amarella L.

Pastures, dry banks and roadsides on the Carboniferous Limestone; rather rare. Native.
1. Hadnock Quarry*, *Charles:* **4.** Wynd Cliff*, *Clark, Shoolbred:* Rogiet, *Hamilton, Bennett, Shoolbred:* Caldicot; Chepstow, *Hamilton:* Llanfihangel Rogiet; St. Brides Netherwent; The Minnetts Common, near Undy*, *Shoolbred:* near Thicket Wood, Rogiet; by Minnetts Wood, *Mrs. Ellis.*

MENYANTHES L.

Menyanthes trifoliata L. Bogbean

Bogs and acid marshes where there is a considerable amount of surface water; rare. Native.
1. Grwyne Fawr Valley; near Pont-y-spig, *Ley,!:* near Penpergwm Station, *Rickards:* base of Coalpit Hill, Fforest, *Charles:* Llanfoist. **2.** Abersychan, *Hamilton:* near Brynmawr. **3.** Near Bettws; Llangybi; Llanusk, *Clark:* near Castell-y-bwch*. **4.** The Glyn, Itton; Coedcae, Shirenewton*, *Shoolbred:* Llwynycelyn, *T. G. Evans.*

POLEMONIACEAE
POLEMONIUM L.
Polemonium caeruleum L.   Jacob's Ladder

Streamside copses; rare. Denizen.
3. Between Llantrisant and Newbridge, *Clark:* Llanfihangel Llantarnam*, *T. H. Thomas.* **4.** Shirenewton, *Clark:* Mounton, *T. G. Evans.*

BORAGINACEAE
CYNOGLOSSUM L.
Cynoglossum officinale L.   Hound's-tongue

Roadsides and quarries; very rare. Native.
1. 'The Leys', Monmouth, *Charles.* **3.** Usk, *Clark:* near Cilfeigan Park*. **4.** Wynd Cliff, *Shoolbred.* **5.** Shingle bank near the Severn between Mathern Oaze and St. Pierre Pill*, *Shoolbred.*

LAPPULA Fabr.

Lappula myosotis Moench
  L. echinata Gilib.
Casual.
1. Field on Kymin Hill, Monmouth*, *Charles.*

144

AMSINCKIA Lehm

Amsinckia lycopsioides Lindl.

Casual.

4. Waste ground, Chepstow, *Lamb.*

SYMPHYTUM L.

Symphytum officinale L.   Comfrey

Hedgebanks, ditches, river- and stream-sides, and waste places; frequent throughout the county. Native.

Both cream- and purple-flowered forms occur.

Symphytum ×uplandicum Nyman   Russian Comfrey
  S. asperum × officinale

Riversides, hedgebanks and waste ground; rare to locally frequent. Denizen.

1. Banks of the River Wye, opposite Dixton Church*; Monmouth Cap*, *Charles:* Pandy*. **3.** Banks of the Afon Lwyd, between Llantarnam and Pontnewydd*; near Castleton*. **4.** Llanvair Discoed*; Mathern*; Newchurch East*, *Shoolbred:* The Coombe, Shirenewton*, *Shoolbred,!:* Tintern; Cuckoo Wood, Llandogo. **5.** Newport Docks*.

(*Symphytum tuberosum* L. recorded by Hamilton from Marshfield is probably an error for the cream-flowered *S. officinale.*)

PENTAGLOTTIS Tausch

Pentaglottis sempervirens (L.) Tausch   Evergreen Alkanet

Roadsides, wood borders and hedgebanks; locally frequent in District **4**, rare elsewhere. Denizen.

1. Kymin Hill, Monmouth*; near Staunton Road, Monmouth*, *Charles:* May Hill, Monmouth, *R. Lewis.* **3.** Pontrhydyrun, near Pontypool*, *T. H. Thomas:* Llwch Wood, Goetre*; between Goetre and Rhyd-y-meirch*. **4.** Near Chepstow, *Clark:* between Bigsweir and Llandogo, *Shoolbred:* Llanvair Discoed*, *Shoolbred,!:* Cuckoo Wood, Llandogo*, *Charles:* Carrow; Shirenewton, *Mrs. Ellis:* Mathern, *Godbey:* near Magor*.

ANCHUSA L.

Anchusa officinalis L.   Alkanet

Waste ground. Garden escape.

3. Usk, *Trans. Woolhope Club.* **4.** Redbrook*, *Charles.*

Anchusa azuria Mill.
  Anchusa italica Retz.

Roadsides; rare. Garden escape.

1. Grosmont, *Charles.* **3.** Near Christchurch*.

BORAGINACEAE

Anchusa arvensis (L.) Bieb.   Small Bugloss
Lycopsis arvensis L.

Cultivated ground and waste places; rare. Colonist.
1. Near the Onen*; *Charles*. **3.** Llanbadoc; Llanusk, *Clark*. **4.** Near the Old Forge, Tintern, *Hamilton, Shoolbred:* Upper Redbrook*, *Charles*.

PULMONARIA L.

Pulmonaria officinalis L.   Lungwort

Woods and hedgebanks; rare. Denizen or garden escape.
1. Beaulieu Wood; Upper Redbrook; Garth Wood*, *Charles*. **3.** Trostra, *Clark*. **4.** Coppice Mawr, near Pandy Mill, Itton*, *Shoolbred:* near Lady Mill, Mounton, *Fraser*.

MYOSOTIS L.

Myosotis scorpioides L.   Water Forget-me-not
M. palustris (L.) Hill

Margins of ponds, reens, streams and canals, and marshes; Common in Districts 1, 3, 4 and 5; rare in District 2. Native.

Myosotis secunda A. Murr.   Creeping Forget-me-not
M. repens auct.

Boggy ground and acid marshes; common in District 2, rather rare or locally frequent elsewhere. Native.
1. Honddu and Grwyne Fawr Valleys, *Ley:* The Ffwddog, *Miss Gee:* near Pont-y-spig. **2.** Common. **4.** Trelleck Bog, *Ley, Shoolbred,!:* St. Arvans*; Penterry*; Chepstow Park; Kilgwrrwg; Tintern*, *Shoolbred:* Pen-y-fan, near Whitebrook*, *Charles:* near Tymawr, Penallt, *Beckerlegge:* Fairoak Pond.

Myosotis caespitosa K. F. Schultz   Tufted Forget-me-not

Margins of ponds, reens, streams and rivers; common. All districts. Native.

Myosotis sylvatica Hoffm.   Wood Forget-me-not

Garden escape. Rare.
1. Railway bank, Pandy*. **4.** Mounton*, *Shoolbred*.

Myosotis arvensis (L.) Hill   Field Forget-me-not

Cultivated ground, hedgebanks, railway banks, walls and open woods; common, especially on the limestone, Rhaetic and New Red Marl. All districts. Native.

146

BORAGINACEAE

Myosotis discolor Pers.   Parti-coloured Scorpion-grass
M. versicolor Sm.

Pastures, meadows and wall tops; locally frequent on the limestone and Old Red Sandstone, absent from District 5. Native.
1. About Llanthony Abbey, *Ley:* near Bell Inn, Skenfrith*, *Charles:* Newton Court Estate, near Monmouth, *R. Lewis:* near Troy Station, Monmouth; Seargent's Grove, near King's Wood*; Triley Bridge*. **2.** Machen, *Hamilton:* Cefn Rhyswg, Abercarn. **3.** Allt-yr-yn; Malpas, *Hamilton:* near Raglan*; near Coed Newydd; Llandegfedd*. **4.** Dewstow, *Hamilton:* near Trelleck, *Watkins:* St. Arvans*; Wentwood Mill*; by Catbrook Lane, Tintern*; near Howick; Itton; near Dinham*; Shirenewton*; Kilgwrrwg, *Shoolbred:* Llandogo*, *Shoolbred, Miss Rickards:* Piercefield, *T. G. Evans:* near Kemeys Inferior.

Myosotis ramosissima Rochel   Early Scorpion-grass
M. collina auct.

Dry banks and quarries on the limestone, and walls; very rare. Native.
1. Walls, Skenfrith*, *Charles.* **4.** Walls, Caerwent*; quarry, Portskewett; between Runstone and Tintern*; near the Old Forge, Tintern, *Shoolbred:* Rogiet, *T. G. Evans.*

LITHOSPERMUM L.

Lithospermum purpurocaeruleum L.   Purple Gromwell

Wood border; very rare. Native.
4. Near Carrow Hill, St. Bride's Netherwent*, *Mrs. Ellis,!.* First seen in 1944; believed to have been destroyed in subsequent forestry operations.

Lithospermum officinale L.   Gromwell

Open woods, thickets, roadsides and railway banks; locally frequent on the limestone, very rare on the Old Red Sandstone. Native.
1. Lady Park Wood; near Hadnock Quarry*; between Monmouth and Hadnock*; near Pandy*, *Charles.* **4.** Between Usk and Llanllywel, *Clark:* near Wynd Cliff, *F. A. Lees,!:* Portskewett; Shirenewton; near Chepstow*, *Shoolbred:* Llandogo*, *Miss Rickards:* between Undy and Rogiet, *Rowlands:* The Minnetts*, *Mrs. Ellis,!:* Carrow Hill; Rogiet.

Lithospermum arvense L.   Corn Gromwell

Cultivated and waste ground; very rare. Colonist.
1. Kymin Hill, Monmouth*, *Charles.* **3.** Cefnila, near Usk, *Clark.* **5.** Newport Docks, *Macqueen.*

ECHIUM L.

Echium vulgare L.   Viper's Bugloss

Roadsides, meadows, waste ground and gravel banks of rivers; very rare. Native.

BORAGINACEAE

1. Govilon; bank of the River Monnow above Pandy, and at Monmouth Cap*, *Charles:* bank of the River Monnow opposite Kenchurch*, *R. Lewis.* **2.** Pontypool, *T. H. Thomas.* **3.** Newport, *Clark:* Usk, *Miss Frederick.* **4.** Near Rogiet*, *Shoolbred:* railway bank, Severn Tunnel, *Mrs. Ellis:* Portskewett, *T. G. Evans.*

CONVOLVULACEAE
CONVOLVULUS L.

Convolvulus arvensis L.   Small Bindweed

Cultivated ground, roadsides, railway banks and waste ground; very common. All districts. Native.

CALYSTEGIA R. Br.

Calystegia sepium (L.) R. Br.   Greater Bindweed

Hedgerows, especially about gardens, and thickets; common in all districts. Native.

Calystegia silvatica (Kit.) Griseb.   Wood Bindweed

Hedgerows; probably commoner than the few records suggest. Denizen. **1.** Llanfoist Golf Course*; Wyesham; Abergavenny; Llantilio Grossenny. **3.** St. Mellons*; Rogerstone; Llangybi. **4.** Near Shirenewton, *Elliott:* Chepstow, *T. G. Evans:* Pwllmeyric.

CUSCUTA L.

Cuscuta epilinum Weihe   Flax Dodder

Parasitic on flax. Alien.
1. Cefnila Farm, near Usk, *Clark*, 1868.

Cuscuta epithymum (L.) L.   Small Dodder

Parasitic on furze, ling and other plants; very rare. Native.
4. Near Rogiet*, *Mrs. Ellis*, 1942.

Cuscuta suaveolens Ser.

Parasitic on various plants. Alien.
2. On *Antirrhinum majus*, Glansychan Park, Abersychan*, *Barker*, 1947.

SOLANACEAE
LYCIUM L.

Lycium barbarum L.   Tea Tree
Lycium halimifolium Mill.

Hedgerows, usually near habitations; rare. Denizen.
1. Wonastow Road, Monmouth*; Rockfield*; Tregare;* Skenfrith; near Chapel Farm, Newton Dixton; Buckholt; Llanfihangel Crucorney; Monmouth Cap*; Cross Ash*, *Charles.* **3.** Llandenny, *J. N. Davies.* **5.** Near Goldcliff.

(Shoolbred's record for *Lycium chinense* Mill. in his *Flora of Chepstow* doubtless refers to this species.)

SOLANACEAE

ATROPA L.

Atropa belladonna L.   Deadly Nightshade

Open woods, roadsides, quarries and railway banks on the limestone; rare. Native.
1. Near Hadnock*; between May Hill Station and Dixton*; Lady Park Wood; Redding's Enclosure*, *Charles:* Staunton Road, near Monmouth, *R. Lewis.* **4.** Live Oaks Quarry, *Price, Shoolbred:* Black Cliff, near Tintern, *Shoolbred:* below Wynd Cliff, *Miss Woodhouse.*

HYOSCYAMUS L.

Hyoscyamus niger L.   Henbane

Waste ground; rare. Alien.
1. Abergavenny*, *T. L. Williams.* **3.** Llwyna Farm, Llantrisant, *Gale:* Chepstow Road, near Christchurch*. **4.** About Chepstow, *E. Lees.* **5.** Near The Severn, Rumney, *Hamilton:* sea wall, Peterstone Wentlloog*, *Watt:* St. Brides Wentlloog, *J. D. Davies:* Sudbrook, *T. G. Evans.*

SOLANUM L.

Solanum dulcamara L.   Woody Nightshade

Hedges, wood borders and banks of streams and reens; common, especially abundant on the marine alluvium. All districts. Native.

Solanum nigrum L.   Black Nightshade

Waste ground, cultivated ground and railway tracks; rather rare. Colonist.
1. Chippenham*; Abergavenny*; Wonastow Road, Monmouth, *Charles:* Dixton Newton, *R. Lewis.* **3.** Usk, *Hamilton.* **4.** Chepstow*, *Clark, Reader, Shoolbred:* Redbrook*, *Charles:* Rogiet, *Mrs. Ellis.* **5.** Newport, *Clark:* Rumney*.

DATURA L.

Datura stramonium L.   Thorn-apple

Waste and cultivated ground; rare. Casual.
1. The Kymin, Monmouth; Chippenham, *Charles.* **3.** Raglan Castle, *E. H. Williams:* Michaelstone y Vedw, *Mrs. Hollings.* **4.** Portskewett, *Shoolbred.* **5.** Marshfield*, *Miss Vachell.*

SCROPHULARIACEAE

VERBASCUM L.

Verbascum thapsus L.   Great Mullein

Dry banks, roadsides, waste ground, walls, railway banks and quarries; frequent throughout the county. Native.

SCROPHULARIACEAE

Verbascum lychnitis L.   White Mullein
Alien.
5. Newport Docks, *Hamilton.*

Verbascum pulverulentum Vill.   Hoary Mullein
Alien.
4. Upper Redbrook*, *Charles,* 1937.

Verbascum nigrum L.   Dark Mullein
Waste ground; rare. Alien.
3. Croesyceiliog, Cwmbran*, *Jeans.* **5.** Newport Docks, *Conway, Hamilton.*

Verbascum blattaria L.   Moth Mullein
Waste ground and refuse tips; rare. Alien.
1. Chippenham Meadow*; Dixton*; near May Hill Station, Monmouth*, *Charles:* Red Hill Farm, Wonastow, *Vernall.*

MISOPATES Raf.

Misopates orontium (L.) Raf.   Weasel-snout
Antirrhinum orontium L.
Cultivated and waste ground; very rare. Colonist.
1. Near Abergavenny*, *Clark,!.* **4.** Caerwent*.

ANTIRRHINUM L.

Antirrhinum majus L.   Snapdragon
Old walls and ruins; rare or locally frequent. Alien.
1. Monmouth, *Charles:* Abergavenny. **2.** Between Cross Keys and Risca.
3. Caerleon*, *Conway:* Penmoyle, near Trostrey†, *Clark:* near Newport, *Hamilton:* near Trostrey Court. **4.** Chepstow*†, *E. Lees, Clark, Hamilton, Shoolbred:* Tintern, *T. Clark, Shoolbred:* Sudbrook, *T. G. Evans:* Magor.

LINARIA Mill.

Linaria purpurea (L.) Mill.   Purple Toadflax
Walls and waste ground; rare. Alien.
1. Kymin Hill, Monmouth*; Wonastow Road, Monmouth*, *Charles.*
3. Bassaleg. **4.** Portskewett*, *Shoolbred.* **5.** Spencer Steel Works*.

Linaria repens (L.) Mill.   Creeping Toadflax
Waste ground, ballast and railway tracks; rare. Alien.
1. Wyesham*, *Charles:* near Pandy Station, *Rees.* **4.** Mathern, *Redgrove:* Pen-y-clawdd, *Miss Davis.* **5.** Newport*, *Conway, Clark, Whitwell:* Llanwern Station, *T. G. Evans.*
×**vulgaris** (*L.* ×*sepium* Allman)
4. Tintern, *Amherst.*

SCROPHULARIACEAE

Linaria vulgaris Mill. Yellow Toadflax

Hedgebanks, roadsides, waste ground, railway banks, walls and limestone quarries; widely distributed throughout the county and common in some areas. All districts. Native.

CHAENORHINUM (DC.) Reichb.

Chaenorhinum minus (L.) Lange Small Toadflax

Railway tracks, quarries, ballast and cultivated ground; locally frequent. Colonist. Less frequent than formerly on railway tracks owing to the use of weed killers.
1. Fields about Llanfair Kilgeddin and Abergavenny, *Purton:* The Onen*; Hendre*; Monmouth; Hadnock*, *Charles:* Llanfoist, *Campbell:* Abergavenny Station. **2.** Rhymney Bridge*; Trefil*; disused railway below Mynydd Dimlaith*. **3.** Usk, *Woolhope Club rep.*, 1867: Raglan district, *Miss Frederick:* near Llanfrechfa*. **4.** Quarries near the Wyndcliff*, *Motley, Shoolbred:* Howick; Llanvair Discoed; Mounton; Severn Tunnel Junction, *Shoolbred:* Mitchel Troy*, *Charles:* Chepstow, *T. G. Evans:* near Thicket Wood, Rogiet*. **5.** Newport Docks, *Hamilton:* Caldicot*; Magor Station*, *Shoolbred,!.*

KICKXIA Dumort.

Kickxia spuria (L.) Dumort. Round-leaved Fluellen
Linaria spuria (L.) Mill.

Cultivated ground and ballast; very rare. Colonist or casual.
4. Near Minnetts Wood, *Mrs. Ellis.* **5.** Newport Docks, *Hamilton.*

Kickxia elatine (L.) Dumort. Sharp-leaved Fluellen
Linaria elatine (L.) Mill.

Cultivated ground, ballast and waste ground; rare. Colonist.
1. Llanthony, *Mathews:* bank of the River Monmow, Monmouth, *Grimes.* The Hendre*, *Charles.* **3.** Rhadyr, Usk, *Clark:* between Bassaleg and Pensylvania; Pen-y-lan Fawr, near Castleton*. **4.** Llanvair Discoed*; near Bigsweir*, *Shoolbred:* Duffield's Farm, Upper Redbrook*, *Charles:* near the Minnetts*; near Rogiet*.

CYMBALARIA Hill

Cymbalaria muralis Gaertn., Mey. & Schorb. Ivy-leaved Toadflax
Linaria cymbalaria (L.) Mill.

Walls; fairly common throughout the county. Denizen.

SCROPHULARIA L.

Scrophularia nodosa L. Knotted Figwort

Woods, shady banks and hedgerows; common. All districts. Native.

SCROPHULARIACEAE

Scrophularia auriculata L. Water Figwort
S. aquatica auct.

Marshes, stream, river and reen sides; common. All districts. Native.

Scrophularia umbrosa Dumort. Winged Figwort
S. alata Gilib.

Riversides; very rare. Native.
4. Bigsweir, *Ley.*

MIMULUS L.

Mimulus guttatus DC. Monkey-flower

Streamsides, marshes and ditches; locally frequent. Denizen.
1. Monmouth; Grwyne Fawr Valley*, *Charles:* Llanthony, *Miss Norman:* Llantilio Crosseny. **2.** Margin of a mountain rill a mile or two from Abergavenny, *Bree:* near Cwmcarn, *Miss Jones:* near Bedwas; Cwm Gwyddon, Abercarn; near Tredegar*. **3.** Llantarnam; Pont-hir, *Hamilton:* Usk, *Miss Frederick, Charles:* Llanbadoc, *Briggs.* **4.** Caerwent, *Hamilton:* near St. Arvans Grange; Tintern; near Wentwood Mill; Llanvair Discoed; Earlswood*, *Shoolbred:* The Combe, Shirenewton*, *Shoolbred, Lannon:* Tintern*, *Perman, T. G. Evans:* New Mills, Penallt*, *Charles:* Whitebrook, *Charles,!:* near Tymawr, Penallt, *Beckerlegge:* St. Brides Netherwent, *Mrs. Ellis:* near Llanbeder*. **5.** Near the dock feeder, Tredegar Park, *Hamilton.*

The Rev. W. T. Bree's record made in 1824 from near Abergavenny was the first for the British Isles.

Mimulus luteus L. Blood-drop Emlets

Shallow streams; very rare. Denizen.
2. Near Abercarn*, *Stumbles.*

Mimulus moschatus Dougl. ex Lindl. Musk

Pond and streamsides; very rare. Denizen.
2. Near Tredegar*, *Burchell.* **4.** Tintern, *Amherst.*

ERINUS L.

Erinus alpinus L.

Naturalized on old walls; rare.
4. Troy House, *Lamb.* Chepstow*, *Hyde.*

DIGITALIS L.

Digitalis purpurea L. Foxglove

Woods and hedgerows; common in all districts except District 5 where it is rather rare. Native.

SCROPHULARIACEAE

VERONICA L.

Veronica beccabunga L. Brooklime

Streamsides, pond margins, marshes and wet woods; common in all districts. Native.

Veronica anagallis-aquatica L. agg. Water Speedwell

Streamsides, reens, marshes and pond margins; locally common, especially on the alluvium of District 5. Native.
1. River Wye, opposite Wyaston Leys, *Charles:* Abergavenny; Llanfoist. **3.** Usk, *Clark:* near Ynys-y-fro, *Hamilton:* Raglan district, *Miss Frederick.* **4.** Langstone, *Hamilton:* Mathern; Tintern; St. Pierre, *Shoolbred:* Trelleck Bog, *Charles:* Mounton; Llwyn-y-celyn, *T. G. Evans.* **5.** Common.

Veronica anagallis-aquatica L. sens. strict.

Pond margins; rare. Native.
3. Near Pont Waun-pwll, Llanddewi Fach*.

Veronica catenata Pennell

Streamsides, reens and marshes; locally common. Native.
1. Hadnock Stream, near Monmouth*; Osbaston*, *Charles.* **4.** St. Pierre Brook*; Mounton Valley*, *Shoolbred:* Gwernesney*, *Charles.* **5.** Common.

Veronica scutellata L. Marsh Speedwell

Bogs, acid marshes and pond margins; rare to locally frequent. Native.
1. Llanfoist*. **2.** Blorenge, *Hamilton:* Pen-y-fan Pond, *McKenzie,!.* **3.** By Ramma Cottage, Usk, *Clark:* near Pen-y-lan, near Castleton; near Pentwyn Farm, Llandegfedd*; near Tynewydd, Henllys*. **4.** Trelleck Bog*, *Watkins, Shoolbred, Charles:* Newchurch West*, *Shoolbred:* Pen-y-fan, near Whitebrook, *Charles:* near Tymawr, Penallt, *Beckerlegge.*

Veronica officinalis L. Common Speedwell

Heaths, rough pastures, hedgebanks and woods; common. All districts. Native.

Veronica montana L. Wood Speedwell

Woods; common on the Old Red Sandstone, frequent on the limestone and Pennant sandstone, Districts 1-4. Native.

Veronica chamaedrys L. Germander Speedwell

Hedgebanks, wood borders and pastures; very common in all districts. Native.

Veronica serpyllifolia L. ssp. **serpyllifolia** Thyme-leaved Speedwell

Pastures, roadsides, wet gravelly places and cultivated ground; common. All districts. Native.

SCROPHULARIACEAE

ssp. **humifusa** (Dickson) Syme

About spring heads in mountain districts; very rare. Native.
1. On the Ffwddog, and on the Monmouthshire faces of the Hatterels, *Ley.*

Veronica arvensis L.　Wall Speedwell

Dry banks, walls, gravelly places and cultivated ground; common. All districts. Native.

Veronica hederifolia L.　Ivy-leaved Speedwell

Cultivated ground, waste ground and sandy loam of hedgebanks; common. All districts. Native.

Veronica persica Poir.　Buxbaum's Speedwell
V. tournefortii Gmel.

Cultivated ground; common. All districts. Colonist.

Veronica polita Fr.　Grey Speedwell

Cultivated ground; common. All districts. Colonist.

Veronica agrestis L.　Field Speedwell

Cultivated ground; rare. Colonist.
2. Pontypool, *T. H. Thomas:* **3.** Risca Road, Newport; Bassaleg, *Hamilton.*
4. Llanyravon, *Conway:* Chepstow*, *Shoolbred.* **5.** Dyffryn, *Hamilton.*
Hamilton's records may be errors for *Veronica polita.*

Veronica filiformis Sm.

Grassy river banks; rare. Denizen.
1. Banks of the River Wye, between Monmouth and Dixton*, *Charles,*
Mrs. Leney. **4.** Banks of the River Wye under Penallt*; between Bigsweir
and Whitebrook*, *Charles.*

PEDICULARIS L.

Pedicularis palustris L.　Red Rattle

Acid marshes and boggy places; rare. Native.
1. Grwyne Fawr Valley; near Pont-y-spig, *Ley.* **2.** Blorenge, *Hamilton:*
Trefil*; near Nant-y-bwch. **3.** Llangybi, *Clark:* near Ynys-y-fro, *Hamilton:*
near Michaelstone y Vedw*; by Coed Mawr, near Rhiwderyn. **4.** Kilgwrrwg Bottom; Catbrook, near Tintern; Tintern Parva, *Shoolbred:*
Llwynycelyn, *T. G. Evans.*

Pedicularis sylvatica L.　Lousewort

Moist pastures, marshes and damp woods; common. All districts. Native.

SCROPHULARIACEAE

RHINANTHUS L.
Rhinanthus minor L.   Yellow Rattle

Pastures, meadows and grassy roadsides. Native.

ssp. **minor.** Common. All districts.

ssp. **stenophyllus** (Schur.) O. Schwartz   Rare.
2. Hafodyrynys, near Crumlin*.   **4.** Shirenewton, *Shoolbred.*

MELAMPYRUM L.
Melampyrum pratense L.   Common Cow-wheat

Woods, heaths and shady river banks; frequent over much of the county but absent from the alluvium of District 5. Native.

(**Melampyrum sylvaticum L.** recorded by Purton from woods about Abergavenny is an undoubted error.)

EUPHRASIA L.
Euphrasia nemorosa (Pers.) Wallr.   Eyebright

Pastures, heaths, limestone quarries and open woods; fairly common on the Old Red Sandstone and limestone of Districts 1, 3 and 4, locally frequent in District 2, absent from District 5. Native.

Euphrasia confusa Pugsley
E. minima auct.

Hillsides in upland areas; rare. Native.
1. Above Llanthony, *Bishop:* Trefil*. **2.** Nantybwch*; Cwm Tysswg*; Mynydd y Garn fawr*.

Euphrasia borealis (Townsend) Wettst.

Hill pastures and meadows, mainly on the Old Red Sandstone; rather rare. Native.
1. Honddu Valley*, *Gough and Sandwith, Charles.* **2.** Cwmcarn, *Miss Jones:* near Beaufort*; Pont-y-spig*; Trefil*. **3.** Near Cwrt Henllys*; near Pant-yr-eos*; near Castell-y-bwch, Henllys*; The Park, Christchurch; Pontypool*. **4.** Earlswood*, *Shoolbred.*

Euphrasia rostkoviana Hayne

Pastures, meadows and heaths; locally frequent. Native.
1. Below Llanthony Abbey, *Gough and Sandwith.* **2.** Near Cwm Lasgarn, Abersychan*. **3.** Cwrt Henllys*; Pant-yr-eos, Henllys*. **4.** Trelleck; Earlswood*; St. Brides Netherwent; The Minnetts; Runstone; between Tintern and Catbrook*; The Ganllwyd*; between The Minnetts and St. Brides*; Bigsweir, *Shoolbred:* Wynd Cliff*, *Shoolbred, R. Lewis:* Troy House, *Riddelsdell:* Newchurch West, *Shoolbred, T. G. Evans:* Mounton*.

SCROPHULARIACEAE

Euphrasia montana Jord.

Moist mountain meadows; rare. Native.
1. Below Tarren-yr-Esgob*. **2.** Cwm Lasgarn, Abersychan*. **3.** Hafodyrynys*.

Euphrasia anglica Pugsl.

Moist meadows; locally frequent. Native.
1. Llanthony, *Ley:* Buckholt wood, near Monmouth*, *Charles.* **4.** Between St. Arvans and the Wynd Cliff†, *Clark:* Shirenewton; Catbrook, *Ley:* between Cockett Inn and Trelleck, *R. Lewis:* Mounton Valley, *Elliott:* Great Barnetts Woods, *T. G. Evans:* between St. Brides and Rogiet*.

ODONTITES Ludw.

Odontites verna (Bell.) Dumort.   Red Bartsia
Pastures, grassy roadsides, open heathy woods and quarries. Native.

ssp. **verna** Common, especially on the limestone and on the marine alluvium of District 5; unrecorded from District 2.

ssp. **serotina** Corbiere
  Bartsia odontites (L.) Huds.

1. Llangattock, *Ley:* Pandy*, *Charles:* Llanvapley. **4.** Pwllmeyric, *Shoolbred:* Tintern, *Watkins.*

OROBANCHACEAE
LATHRAEA L.

Lathraea squamaria L.   Toothwort

Parasitic on the roots of various trees. Woods; locally frequent. Native.
1. Honddu Valley, *Bull:* near Highmeadow Siding*; Highmeadow Woods; Hadnock Wood*; Lady Park Wood*; Redding's Enclosure*, *Charles.* **4.** Between Chepstow and Wynd Cliff, *F. A. Lees:* near Chepstow, *Hamilton:* Black Cliff, near Tintern; Castle Woods, Chepstow*; Fryth Wood, *Shoolbred:* between Penterry Farm and Tintern*, *Charles:* The Coombe, Shirenewton, *Lannon:* Piercefield Cliffs, *T. G. Evans:* near Tintern*.

OROBANCHE L.

Orobanche purpurea Jacq.   Purple Broomrape

Parasitic on yarrow; very rare. Native.
4. A mile or so south-west of Chepstow, *Hort:* Mathern Parish*, *Mrs. Francis:* Langham's Farm, Chepstow*, *Shoolbred.*

Dr. Hort's record, made in 1852, may refer to the same locality in which Mrs. Francis collected the plant in 1912.

OROBANCHACEAE

Orobanche rapum-genistae Thuill.   Greater Broomrape
O. major auct.
Parasitic on gorse and broom; very rare. Native.
1. Buckholt Wood*, *Charles*. **3.** Bettws Newydd, *Clark*. **4.** Near Chepstow*, *Mrs. Welch*.
Walford's record from Chepstow Castle, in his *Scientific Tourist*, 1, 1818, probably refers to *Orobanche hederae*.

Orobanche minor Sm.   Lesser Broomrape
Parasitic on clover and other plants; rare. Native.
1. Hadnock Farm, near Monmouth*, *Charles*. **3.** Rhadyr, near Usk, *Clark:* Usk*; Coed-y-paen*, *Rickards:* near Christchurch*. **4.** Llandogo; Bigsweir, *Watkins:* Mathern*, *Mrs. Francis, Shoolbred:* Trelenny; near Chepstow, *Shoolbred:* Rogiet; The Minnetts, *Mrs. Ellis.* Caldicot*, *Rees.*

Orobanche hederae Duby   Ivy Broomrape
Parasitic on ivy; rare to locally common. Native.
1. Monmouth Castle, *Woods:* near Ysgyryd Fawr, *E. Lees:* about Llanthony Abbey, *Ball.* **3.** Raglan Castle, *Trapnell.* **4.** Chepstow Castle*, *E. Lees, Shoolbred:* Wynd Cliff, *Trans. Worc. Nat. Club*, 1896: plentiful about limestone cliffs by the Wye, *Shoolbred:* Portwall School walls, *T. G. Evans.*

LENTIBULARIACEAE
PINGUICULA L.

Pinguicula vulgaris L.   Butterwort
Bogs and acid marshes; frequent in the Honddu and Grwyne Fawr Valleys, very rare elsewhere. Native.
1. Honddu and Grwyne Fawr Valleys; the Hatterels and the Ffwddog, *Ley:* about Llanthony*, *T. H. Thomas, Hamilton, Charles.* **2.** Varteg†; Pontypool, *Clark:* Trefil*. **3.** near Castlell-y-bwch, Henllys*, *Sculthorpe,!.* **4.** Trelleck Bog*, *Hamilton, Shoolbred,!.*

UTRICULARIA L.

Utricularia neglecta Lehm.   Bladderwort
U. major auct.
Ditches and reens; very rare. Native.
4. Wye Valley, near Chepstow, *Hamilton* (as *U. vulgaris*). **5.** Reens between Lower End, Magor and Llandevenny*, *Dr. Clarke:* Magor*, *Shoolbred:* Magor-Llanwern, 1957-59, not seen in the 1960's following spraying with herbicides, *T. G. Evans.*

(**Ultricularia vulgaris** L. The record in Shoolbred's *Flora of Chepstow* is referable to *U. neglecta.*)

157

VERBENACEAE

VERBENA L.

Verbena officinalis L.   Vervain

Roadsides and waste ground; locally frequent. Native.
1. Near Wyesham, *Charles,!:* Abergavenny; Llanfoist*. **2.** Abercarn, *McKenzie:* Risca. **3.** Cardiff Road, near Tredegar Park; Christchurch, *Hamilton:* Raglan district, *Miss Frederick:* Llanrumney*. **4.** Bishpool, *Hamilton:* Wynd Cliff; Shirenewton, *Shoolbred:* Tintern, *Amherst:* Redbrook*, *Charles:* Llanvair Discoed, *Lannon:* Rogiet, *Mrs Ellis:* Chepstow, *Shoolbred, T. G. Evans:* Mathern Mill, *T. G. Evans:* gravel pit, Black Cliff, near Tintern.

LABIATAE

MENTHA L.

Mentha pulegium L.   Pennyroyal

Marshy places; very rare. Native.
4. Rogiet, *Mrs. Ellis.*

Mentha arvensis L.   Field Mint

Cultivated ground, damp roadsides, marshy ground and waste ground; common. All districts. Native.

Mentha ×gentilis L.   Fringed Mint
  M. arvensis × spicata.

River banks; rare. Denizen.
1. Opposite The Bibblings*; bank of the River Monnow, near Monmouth Cap*, *Charles:* Hadnock, *R. Lewis.* **4.** Near Llandogo*, *Shoolbred:* near Penallt*, *Charles.*

Mentha aquatica L.   Water Mint

By rivers, streams and ponds, and marshes; common. All districts. Native.

Mentha ×verticillata L.   Whorled Mint
  M. aquatica × arvensis

River sides and marshy places; common in Districts **1** and **4**, occasional elsewhere. Native.

Mentha ×smithiana R. A. Graham   Red Mint
  M. aquatica × arvensis × spicata; M. rubra Sm. non Mill.

River banks and streamsides; rare. Denizen.
1. Bigsweir, *Ley:* Osbaston*, *Charles:* near Llanfoist. **3.** By the Afon Lwyd, between Llanvihangel Llantarnam and Pontnewydd; Llangattock juxta Caerleon. **4.** Mounton, *Shoolbred:* between Whitebrook and Penallt*; Botany Bay, Tintern*, *Charles.*

Mentha × **piperita** L.   Peppermint
M. aquatica × *spicata*

River banks, streamsides and ditches; rare. Denizen.
1. Near Monmouth Cap*; near Hadnock Sidings*, *Charles:* Govilon;
Osbaston*. **3.** Near Llanbadoc, *Clark.* **4.** Mounton; Tintern; Penterry*,
Shoolbred: Botany Bay, Tintern, *R. Lewis.* **5.** Near Pye Corner, Nash*.

Mentha spicata L.   Spear Mint

Waste ground, roadsides and river banks; rare. Naturalized garden escape.
1. Abergavenny, *E. Lees:* about Llanthony Abbey, *Ball:* banks of the
Honddu below Cwmyoy, *Ley:* Monmouth*; Hadnock*, *Charles:* **2.** By
rivulets ascending the Blorenge, *E. Lees:* Bedwas. **3.** Between Llanfihangel
Llantarnam and Pontnewydd.

Mentha longifolia (L.) Huds.   Horse Mint

River banks, marshy places and wet roadsides; frequent in Districts 1
and 4, rare elsewhere. Doubtfully native.
1. About Llanthony Abbey, *Ball:*Cwmyoy, *Ley:* near Greenmoors, Llan-
ddewi Ysgyryd*, *Willan:* near Hadnock Quarry*; Monmouth;Osbaston*;
Upper Redbrook*; Skenfrith; by the River Monnow, Monmouth Cap*,
Charles: near Llantillio Crossenny*, *W. Nelmes:* near Llanfoist*. **2.** The
Blorenge, *E. Lees.* **3.** Raglan, *Miss Frederick:* Llanbadoc*, *Harrison:*
Llangattock juxta Caerleon; near Pentwyn Farm, Llanddewi Fach*; Plas
Machen. **4.** Mounton Valley; Old Forge, Tintern*; The Coombe, Llanvair
Discoed*, *Shoolbred:* Penallt*; near Whitebrook*, *Charles.* **5.** Near the
Belt, Dyffryn; St. Brides Wentlloog; Lliswerry; Nash, *Hamilton.*

× **rotundifolia** (*M.* × *niliaca* Juss ex Jacq.; *M. alopecuroides* Hull).
3. Malpas, *Hopkinson:* Llanbadoc*, *Harrison.* **4.** The Coombe, Llanvair
Discoed, *Ley.* **5.** Marshfield*, *Miss Vachell.*

Mentha rotundifolia (L.) Huds.   Round-leaved Mint

Roadsides, waste ground, river banks and streamsides. Widespread but
not common. Doubtfully native.
1. Llanfair Kilgeddin, *Purton:* near Monmouth*, *Purton, Woods, Charles:*
Hadnock*, *Charles:* between Monmouth and Watery Lane*. **2.** Cwm-
carn*, *McKenzie:* New Tredegar, *J. W. Thomas:* near Maes-y-cwmmer*.
3. Near Pontnewydd, *Conway:* between Llanfihangel Llantarnam and
Pontnewydd; Pont-hir; Llangattock; Llanbadoc. **4.** Tintern, *Lightfoot,
Miss Roper:* Llanvair Discoed, *Forster:* Wentwood, *Hamilton:* Chepstow;
Wynd Cliff; between Tintern and Bigsweir; Trelleck; Mounton*; The
Coombe, Llanvair Discoed*; Itton*, *Shoolbred:* Whitebrook Valley*;
between Whitebrook and Penallt*, *Charles:* Rogiet, *Mrs. Ellis:* near
Tintern, *T. G. Evans:* Lady Mill, Mounton*. **5.** Near Rumney*.

First recorded by John Lightfoot in 1773 from Tintern.

159

LABIATAE

LYCOPUS L.

Lycopus europaeus L.   Gipsywort

Marshes and by rivers and streams; frequent in all districts. Native.

ORIGANUM L.

Origanum vulgare L.   Marjoram

Hedgebanks, open woods and roadsides, chiefly on the limestone; locally frequent to locally common. Native.
1. About Clytha, *Purton:* Monmouth; Lady Park Wood, *Charles:* Lord's Grove. **2.** Risca; Crosskeys*. **3.** Usk, *Purton:* Rhiwderyn, *Hamilton:* Raglan district, *Miss Frederick:* between Castleton and Michaelstone y Vedw*. **4.** Common. **5.** Newport Docks*, *Macqueen.*

THYMUS L.

Thymus pulegioides L.   Larger Wild Thyme
 T. ovatus Mill.; *T. chamaedrys* Fr.

Pastures, heaths and dry banks, chiefly on the limestone; rare. Native.
1. Llangattock Vibon Avel; upper part of the Grwyne Fawr Valley, *Ley.* **2.** Near Craig Gwyn, Abercarn*; Craig yr Hafod, near Blaenavon*. **4.** Chepstow, *Monington:* below Raven's Nest Wood, near Tintern*; Earlswood, *Shoolbred:* by Slade Wood, Rogiet*; Penallt Common*.

Thymus drucei Ronn.   Lesser Wild Thyme

Heaths, pastures, and dry banks, chiefly on the limestone; locally common. Native.
1. The Hendre, near Monmouth, *Hamilton:* Monmouth, *Charles:* Pont-y-spig. **2.** Mynydd Maen, *Briggs:* The Blorenge; Machen; Trefil; Mynydd Machen; near Ebbw Vale; near Aberbeeg. **3.** Earlswood, *Clark:* near Coed Newydd, Trostrey. **4.** Common.

CALAMINTHA Mill.

Calamintha ascendens Jord.   Calamint

Hedgebanks and dry grassy places, chiefly on the limestone; rare. Native.
1. Opposite the Bibblings*; Redding's Enclosure*; White Castle, near Llantilio Crossenny*; Lady Park Wood, *Charles:* Grosmont Castle, *Miss Newton.* **3.** Pant-yr-heol, Henllys. **4.** Ifton; Wynd Cliff; Caldicot, *Hamilton:* Portskewett*; Llanvair Discoed, *Shoolbred.*

(The record for *Calamintha nepeta* (L.) Savi, in Watson's *Topographical Botany*, is probably an error.)

ACINOS Mill.

Acinos arvensis (Lam.) Dandy   Basil Thyme

Roadsides; very rare. Native.
1. Lady Park Wood, *Charles.* **4.** Near Magor, *Dr. Clarke:* Grey Hill, Wentwood, *Mrs. Ellis.*

LABIATAE

CLINOPODIUM L.

Clinopodium vulgare L.   Wild Basil

Hedgebanks, wood borders and rough pastures; common. All districts. Native.

MELISSA L.

Melissa officinalis L.   Balm

Hedgebanks, roadsides and waste ground, usually near houses; rare to locally frequent. Denizen.

1. Govilon; Redbrook*; Norton*; Skenfrith, *Charles*. **3.** Between Cilfeigan and Coed-y-paen*; between Castleton and St. Mellons. **4.** Chepstow, *Hort:* St. Arvans; near The Innage, Mathern*; near Pill Farm, Magor, *Shoolbred:* Tintern*, *Druce, Shoolbred:* near Penallt Halt*, *Charles:* between Pen-y-clawdd and Dingestow, *Sandwith:* Five Lanes, Caerwent, *Mrs. Ellis:* Redbrook*; Bishton; Llandogo. **5.** Coast road, near Rumney*.

SALVIA L.

Salvia verticillata L.   Whorled Sage

Alien.

5. Newport Docks, 1952*, *Maqueen*.

Salvia pratensis L.   Meadow Sage.

Pastures; very rare. Native.

4. Rogiet*, *Miss Vachell, Shoolbred, Hamilton,!:* Caldicot, *Hamilton.*

An anonymous MS record from Abergavenny Castle in a copy of Mavor's *Botanical Pocket Book* for 1807 probably referred to a garden relic.

Salvia horminioides Pourr.   Wild Sage, Clary

  S. verbenaca auct.

Roadsides and river banks; rare. Native.

5. Banks of the River Usk, Newport, *Clark*. **4.** Chepstow, *Clark:* between Chepstow and St. Arvans*, *Shoolbred:* Wye Valley, near Chepstow, *Hamilton:* Rogiet, *T. G. Evans.*

PRUNELLA L.

Prunella vulgaris L.   Selfheal

Pastures, meadows and grassy roadsides; common in all districts. Native.

BETONICA L.

Betonica officinalis L.   Wood Betony

  Stachys officinalis (L.) Trev.

Hedgebanks, woods, heaths, grassy roadsides and pastures; common. All districts. Native.

161

LABIATAE

STACHYS L.

Stachys arvensis (L.) L.   Corn Woundwort

Cultivated ground; locally common in Districts 1, 3 and 4, chiefly on the Old Red Sandstone; rare in District 2 and restricted to the south-east; absent from District 5. Native.

Stachys palustris L.   Marsh Woundwort

Sides of rivers, canals and streams, and marshy places; frequent. All districts. Native.

×**sylvatica** (*S.* × *ambigua* Sm.)

1. Llantilio Crossenny*, *E. Nelmes*. **2.** Bedwas. **3.** Near Pontnewydd*; Goytre. **4.** Trelleck, *Ley:* St. Arvans; Shirenewton, *Shoolbred.*

Stachys sylvatica L.   Hedge Woundwort

Hedgebanks and woods; very common in all districts. Native.

BALLOTA L.

Ballota nigra L.   Black Horehound

Roadsides and waste places, usually near houses and farm buildings; not common but widely distributed in Districts 1, 3 and 4, rare in District 5 and unrecorded from District 2. A doubtful native.

LAMIASTRUM Heist. ex Fabr.

Lamiastrum galeobdolon (L.) Ehrend. & Polatsch.

  Galeobdolon luteum Huds.   Yellow Archangel

Hedgebanks and woods; common in all districts. Native.

LAMIUM L.

Lamium amplexicaule L.   Henbit

Cultivated ground; very rare. Colonist.

3. Malpas, *Hamilton:* **5.** Dyffryn; Pencarn; St. Brides Road, *Hamilton.*

Lamium hybridum Vill.   Cut-leaved Deadnettle

Cultivated ground; very rare. Colonist or casual.

3. Near Pontnewydd, *Conway.*

Lamium purpureum L.   Red Deadnettle

Cultivated ground, roadsides and waste places; common. All districts. Native.

Lamium album L.   White Deadnettle

Roadsides, hedgebanks and waste places, usually near buildings; not very common but widely distributed in all districts. Denizen.

Lamium maculatum L.   Spotted Deadnettle

Hedgebanks; very rare. Garden escape.

1. Llanfihangel, *Ley:* Monmouth Cap*, *Charles.*

LEONURUS L.

Leonurus cardiaca L.   Motherwort

Roadsides and waste ground; rare. Alien.

1. Abergavenny district, *White.* **2.** Near Gelligroes Mill, *McKenzie.* **3.** Christchurch, *Lightfoot:* Lower Machen*, *Perman.* **4.** Tintern*, *Shoolbred:* Pen-y-clawdd*, *Miss M. Parry.*

First recorded by John Lightfoot in 1773.

GALEOPSIS L.

Galeopsis angustifolia Ehrh. ex Hoffm.   Red Hempnettle
   G. ladanum auct.

Cultivated and waste ground; very rare. Colonist or casual.

3. St. Woolas, *Hamilton.* **5.** Forge Lane, Dyffryn, *Hamilton.*

(**Galeopsis speciosa** Mill. recorded by Conway in Watson's *New Botanists' Guide*, 1837, is probably an error for *G. tetrahit.*)

Galeopsis tetrahit L. *sens. lat.*   Hempnettle

Hedgebanks, wood borders, roadsides and cultivated ground. Native.

Galeopsis tetrahit L. *sens. strict.* Frequent. All districts.

Galeopsis bifida Boenn.   Rare.
   G. tetrahit var. bifida (Boenn.) Lej. & Court
4. The Glyn, Itton*, *Shoolbred.*

NEPETA L.

Nepeta cataria L.   Catmint

Hedgebanks, roadsides and waste ground; rare. Denizen.
2. Rhymney Valley, *Hamilton.* **4.** Troy House, Monmouth, *Ley:* Caerwent*, *Shoolbred, Hamilton.* **5.** Ballast, Newport, *Clark.*

GLECHOMA L.

Glechoma hederacea L.   Gound Ivy
   Nepeta hederacea (L.) Trev.

Hedgebanks and woods; very common in all districts. Native.

163

LABIATAE

MARRUBIUM L.

Marrubium vulgare L.   White Horehound

Roadsides and waste ground; very rare. Denizen.
1. Near the River Wye, above Monmouth, *Trapnell*. **4.** Castle grounds, Chepstow, *Hamilton*. **5.** Coedkernew, *Hamilton*.

SCUTELLARIA L.

Scutellaria galericulata L.   Skull-cap

River and canal banks, streamsides, reens and marshes; locally frequent. Native.
1. Pont-y-spig*, *Ley,!:* Llanfoist. **2.** Cwmbran*, *T. H. Thomas:* Abercarn, *McKenzie:* Risca*. **3.** Banks of the Olway Brook, near Usk, *Hamilton:* Malpas, *Hamilton, Hopkinson,!:* Usk, *Miss Frederick:* Allt-yr-yn; Ynys-y-fro; near Plas Machen*; near Castell-y-bwch. **4.** The Glyn, Itton*; Newchurch, *Shoolbred:* Whitebrook; Redbrook*; Trelleck Bog*, *Charles:* Llwyn-y-celyn, *T. G. Evans*. **5.** Tredegar Park, *Hamilton:* Peterstone Went-lloog*; Coedkernew*; Whitson.

Scutellaria minor Huds.   Lesser Skull-cap

Bogs and acid marshes; locally frequent. Native.
2. Nant Gallon Wood, near Pontypool*, *T. H. Thomas:* Mynydd Dim-laith*; below Twmbarlwm. **3.** Garw Wood, near Pontnewydd*, *Conway,!:* near Rhadyr, *Clark:* near Foxwood; near Ynys-y-fro, *Hamilton:* near Pentwyn Farm, Llanddewi Fach*. **4.** Near Pen-y-cae-mawr; near Earls-wood, *Clark:* Mounton; Barbadoes Hill; Fairoak Pond, near Tintern*; The Narth, *Shoolbred:* Trelleck Bog, *Shoolbred,!:* Pen-y-fan, near White-brook*; Catbrook*, *Charles:* Whiteley Common.

TEUCRIUM L.

Teucrium scorodonia L.   Wood Sage

Hedgebanks and woods; common. All districts. Native.

Teucrium flavum L.

Garden escape.
4. Pwllmeyric*, *Shoolbred*.
Recorded by Shoolbred in his *Flora of Chepstow* as *Teucrium chamaedrys*.

AJUGA L.

Ajuga reptans L.   Bugle

Woods, hedgebanks and damp pastures; common. All districts. Native.

PLANTAGINACEAE
PLANTAGO L.
Plantago major L.   Greater Plantain

Pastures, roadsides, cart tracks, waste and cultivated ground; very common in all districts. Native.

Plantago media L.   Hoary Plantain

Pastures, grassy roadsides and banks; locally common on the limestone. Native.
1. Hadnock*; near May Hill Station, Monmouth*; near Fiddler's Elbow, Monmouth*; Porth-y-gwaelod Farm, Rockfield*, *Charles*. **2.** Near Pont-llan-fraith, Abercarn, *McKenzie*. **4.** Chepstow*, *Conway, Clark, Hamilton, Shoolbred:* Ifton; Dewstow, *Hamilton:* Penallt*, *Charles:* near Carrow Hill.

Plantago lanceolata L.   Ribwort Plantain

Pastures, grassy roadsides and banks, waste ground and walls; very common in all districts. Native.

Plantago maritima L.   Sea Plantain

Salt marshes and banks of tidal rivers; common along the Severn and the river mouths. District 5. Native.

First recorded by John Lightfoot in 1773.

Plantago coronopus L.   Buck's-horn Plantain

Tidal river banks and grassy places near the sea; common along the Severn and the river mouths. District 5. Native.

Plantago indica L.

Casual.
5. Garden weed, Well House, Marshfield*, *Miss Stratton*, 1957.

LITTORELLA Berg.
Littorella uniflora (L.) Aschers.   Shoreweed

Margins of lakes; rare. Native.
2. Pen-y-fan Pond, near Oakdale*. **3.** Wentwood Reservoir*, *Mrs. Parris*.

CAMPANULACEAE
WAHLENBERGIA Schrad.
Wahlenbergia hederacea (L.) Reichb.   Ivy-leaved Bellflower

Bogs and damp peaty places; locally frequent. Native.
2. Abersychan, *Clark:* near the mouth and the headwaters of Nant

CAMPANULACEAE

Gwyddon, Abercarn; Sirhowy Valley, *McKenzie:* near Mynyddislwyn, *Mrs. Leney:* Mynydd Dimlaith*; Bedwas; below Twmbarlwm, Risca; Cwm Big, Aberbeeg*. **4.** Shirenewton, *Clark:* Llanvair Discoed*; near Lower House, Earlswood; by Nant Trogi, near Cribba Mill, Wentwood, *Shoolbred.*

CAMPANULA L.

Campanula latifolia L. Giant Bellflower

Woods, thickets, hedgebanks, railway banks and shady river banks; locally frequent. Native.
1. Abergavenny district, *White:* banks of the River Monnow, *Woods:* near Grosmont; Hadnock; between Dixton and Hadnock; Monmouth Cap*; Lady Park Wood; banks of the River Wye, Monmouth*, *Charles:* near Great Goytre, Grosmont. **3.** Near Llanbadoc†; Upper Llancayo†; *Clark:* Usk, *Clark, Mrs. Scholberg:* Raglan district, *Miss Frederick:* between Llantarnam and Pontnewydd*. **4.** Itton*; near Tintern; Wynd Cliff; Bigsweir, *Shoolbred:* Kilgwrrwg, *Redgrove:* Redbrook, *McLean:* near Llanishen, Penallt*; between Monmouth and Redbrook, *Charles:* Pen-y-clawdd, *Sandwith.*

Campanula trachelium L. Nettle-leaved Bellflower

Woods, hedgebanks and river banks; rare except in District 1 where it is locally frequent. Native.
1. Near Clytha*, *Shoolbred:* Near Hadnock*; by the River Wye, near Martin's Pool; Lilyrock Wood; Fiddler's Elbow, Monmouth*; Redding's Enclosure; between Llanvapley and Abergavenny*; near May Hill Station, Monmouth; Lady Park Wood; Buckholt Wood, *Charles:* Tregare, *Rees.* **3.** Between Usk and Abergavenny, *Clark:* Raglan district, *Miss Frederick:* Usk, *Mrs. Ellis.* **4.** Tintern, *Shoolbred:* Penallt*, *Charles,!.*

Campanula rotundifolia L. Harebell

Meadows, hedgebanks, heaths, wood borders and grassy roadsides; rather thinly scattered throughout the county but common in some areas; absent from District 5. Native.

Campanula patula L. Spreading Bellflower

Woods, hedgebanks and railway banks; locally frequent. Native.
1. Between Pandy Station and the Hatterels, *E. Lees:* near Skenfrith, *C. Parkinson:* Beaulieu Wood*; near the War Fields Cottage, Coleford Road; Hadnock Wood*; Pritchard's Hill; between Monmouth and Staunton*; Redding's Enclosure; above the Coleford Road, near Monmouth, *Charles:* Priory Grove Wood, Monmouth, *R. Lewis.* **3.** Coed-y-paen, *Conway:* near Llangybi, *Forster:* Cefnila, *Rickards,!:* near Trostrey Common, *Arthur.* **4.** Near Llanllywel, *Forster:* Bigsweir; between Tintern

CAMPANULACEAE

and Trelleck, *Watkins:* between Bigsweir and Monmouth, *Bryan:* Tintern, *Ley, Hamilton:* between Wyesham and Redbrook; near Whitebrook*, *Charles.*

Campanula medium L.   Canterbury Bell
Garden Escape.
3. Hedgebank, Pontnewydd*.

JASIONE L.

Jasione montana L.   Sheep's-bit Scabious
Heaths, banks and woods on siliceous soils; locally common. Native.
1. Gaer Hill, *Bull & Watkins:* Honddu and Grwyne Fawr Valleys, *Ley.*
2. Common. 4. Tintern*; between Tintern and Trelleck*; between Tintern and the Narth; Wyes Wood; Earlswood; near Nine Wells Farm, *Shoolbred:* Cleddon*; Trelleck*; Maryland*, *Charles:* near Lydart, Penallt, *Beckerlegge.*

RUBIACEAE

SHERARDIA L.

Sherardia arvensis L.   Field Madder
Cultivated ground, walls and sea walls; common. Districts 1, 3-5. Colonist.

ASPERULA L.

Asperula arvensis L.
Casual.
1. Site of a poultry run, Kymin Hill, Monmouth*, *Charles.*

CRUCIATA Mill.

Cruciata laevipes Opiz   Crosswort
Galium cruciata (L.) Scop.
Hedgebanks, wood borders, roadside banks and pastures; common on basic soils. Districts 1-4. Native.

GALIUM L.

Galium odoratum (L.) Scop.   Sweet Woodruff
Asperula odorata L.
Woods and shady hedgebanks; locally common. Native.
1. Common. 2. Pontypool, *T. H. Thomas:* Lasgarn Wood, Abersychan; between Abergavenny and Blaenavon. 3. St. Julians, *Hamilton:* Raglan district, *Miss Frederick:* St. Mellons. 4. Common.

Galium mollugo L.   Hedge Bedstraw
Hedgebanks, bushy places, railway banks and rough pastures. Native.

167

RUBIACEAE

ssp. **mollugo** Frequent to locally common. Districts 1, 3-5.

ssp. **erectum** Syme Rare and restricted to the limestone.
4. Wynd Cliff, *Marshall:* The Barnetts*; Kilgwrrwg; between Tintern and the Wynd Cliff*, *Shoolbred.*

Galium verum L.   Lady's Bedstraw
Pastures, grassy roadsides and dry banks; locally common. Native.
1. About Monmouth; Hadnock*; *Charles:* between Monmouth and Drewen Cottages; Tarren yr Esgob. **2.** Mynydd Maen, *Briggs.* **3.** Wern Fawr Wood, near Nant-y-deri, *Briggs.* **4.** Common. **5.** Severn Tunnel, *Hamilton:* Sudbrook, *T. G. Evans.*

Galium saxatile L.   Heath Bedstraw
Heaths, moorlands, pastures, stony places and walls; common, especially in the upland districts. Districts 1-4. Native.

Galium pumilum Murr.   Mountain Bedstraw
  G. sylvestre Poll.
Pastures and rocky places; very rare. Native.
2. Trefil*.

Galium palustre L. *sens. lat.*   Marsh Bedstraw
Marshes, bogs and by reens and streams; common. All districts. Native
ssp. **palustre** Common.
ssp. **elongatum** (C. Presl) Lange
3. Canal near Bettws Lane, Newport. **5.** Magor*, *Shoolbred:* Rumney*.

Galium uliginosum L.   Bog Bedstraw
Bogs and peaty marshes; rare. Native.
1. Grwyne Fawr Valley; near Pont-y-spig*, *Ley,!.* **3.** Near Foxwood *Hamilton:* Garw Wood, near Pontnewydd; near Pentwyn Farm, Llanddewi Fach*; near Castell-y-bwch*. **4.** Near Fair Oak Pond, Tintern*; Coed Cae, near Shirenewton*, *Shoolbred:* near Langstone, *Hamilton:* Trelleck Bog*, *Charles:* by the Virtuous Well, Trelleck*.

Galium tricornutum Dandy   Rough-fruited Goosegrass
  G. tricorne Stokes *pro parte*
Cultivated and waste ground; rare. Casual.
1. Kymin Hill, Monmouth*, *Charles.* **4.** Near Llanmelin*, *Shoolbred.*

Galium aparine L   Goosegrass, Cleavers
Hedgebanks and wood borders; very common. All districts. Native.

RUBIACEAE

RUBIA L.

Rubia peregrina L.   Wild Madder

Hedges and woods on the limestone; locally frequent. Native.
1. Lady Park Wood*, *Charles, Young.* **4.** Near Crick, *Ray:* Piercefield Woods, *Lightfoot, Cullum:* Wynd Cliff†, *Motley, Clark:* Black Cliff Wood, *Gissing,!:* between Shirenewton and Crossway Green Gate†; Penmoyle†; Cophill, *Clarke:* The Coombe, Llanvair Discoed, *Shoolbred:* near Chepstow; Chepstow Castle, *Hamilton:* The Minnetts, *Mrs. Ellis.*

First recorded by John Ray in 1662.

CAPRIFOLIACEAE

SAMBUCUS L.

Sambucus ebulus L.   Danewort

Roadsides and about ruins; rare. Denizen.
1. Abergavenny, *E. Lees:* foot of the Deri near Llanddewi Ysgyryd, *Ley:* near Llanthony Court,* *Bailey:* between Llanddewi Ysgyryd and Llanvetherine, *Sandwith.* **2.** Between Newbridge and Chapel of Ease, *McKenzie.* **3.** Raglan Castle; Caerleon, *Donovan:* Llanfair Cilgedin, *Rickards:* near Pontypool*, *McKenzie.* **4.** Caerwent*, *Manby, Donovan, Shoolbred, Hamilton, Redgrove:* Llanwern, *Hamilton.*

Sambucus nigra L.   Elder

Hedges and woods; very common. All districts. Native.

VIBURNUM L.

Viburnum lantana L.   Wayfaring Tree

Hedges and woods on calcareous soils; locally common. Native.
1. Lilyrock Wood; Redding's Enclosure*; Lady Park Wood, *Charles,!:* Raglan Road, near Dingestow*, *Gwilliam.* **2.** Llanover Woods, *Hamilton.* **3.** St. Julian's Wood, Christchurch, *Hamilton.* **4.** Wynd Cliff, *Clark:* about Chepstow; Itton; Barnett Woods*; Piercefield*, *Shoolbred:* Upper Redbrook*, *Charles:* The Minnetts; Black Cliff Wood; Lady Hill, Newport*; between Bishton and Llanmartin.

Viburnum opulus L.   Guelder Rose

Hedges and woods; common. All districts. Native.

SYMPHORICARPOS Duham.

Symphoricarpos rivularis Suksd.   Snowberry
   S. racemosus auct.

Hedges, usually near houses; frequent in Districts 1, 3 and 4, rare in 2 and 5. Denizen.

CAPRIFOLIACEAE

1. Near Llanthony Abbey, *Ley:* Kymin Hill*; Dixton*; Talycoed*; Llanvapley*; St. Maughan's Green*, *Charles:* Monmouth; Pandy; Abergavenny*. **2.** Near Llwyn-llynfa, Bedwas. **3.** By Tredegar Park; Malpas Road, Newport; The Plantation, Croesyceiliog; Christchurch; Bassaleg*. **4.** Between Mounton and the Grondra*, *Shoolbred:* Milton; Pwlpan; near Kemeys Inferior; Kites Bushes, Mounton*. **5.** Near St. Brides Wentlloog.

LONICERA L.

Lonicera periclymenum L. Honeysuckle

Hedges and wood borders; common. All districts. Native.

Lonicera nitida Wils.

Denizen.

4. Piercefield Woods, 1968.

LEYCESTERIA Wall.

Leycesteria formosa Wall. Himalayan Honeysuckle

Denizen.

4. Cuckoo Wood, near Llandogo, 1944*.

ADOXACEAE

ADOXA L.

Adoxa moschatellina L. Moschatel

Shady hedgebanks and woods, especially on rich loamy soil; frequent to common. All districts. Native.

VALERIANACEAE

VALERIANELLA Mill.

Valerianella locusta (L.) Betcke Common Corn-salad
  V. olitoria (L.) Poll.

Cultivated ground, walls and railway tracks; rare to locally frequent. Native.

1. Near Pont-y-spig, *Woolhope Club Trans.* 1885. **2.** Near the Folly, Pontypool*, *T. H. Thomas.* **3.** Caerleon; near Pontsampit Bridge, Usk†, *Clark.* **4.** Chepstow*, *Shoolbred:* Llandogo*, *Shoolbred,!:* near Severn Tunnel; Portskewett, *Hamilton:* near Penallt*; Troy House*; near Bigsweir*; Whitebrook*, *Charles:* Highmoor Hill; Brockwells, between Dewstow and Caerwent, *Mrs. Ellis:* Penhow; Tintern*.

Valerianella carinata Loisel. Keel-fruited Corn-salad

Banks and walls; rare. Native.

VALERIANACEAE

1. Abergavenny, *White:* Llanover*, *Freer.* 4. Tintern, Chepstow, *Ley:* Mounton, *Shoolbred.*

Shoolbred's record from near St. Lawrence, Chepstow is erroneous; the specimen in his herbarium (N.M.W.) is *V. locusta.*

Valerianella rimosa Bast.   Broad-fruited Corn-salad

Cultivated ground; very rare. Colonist.
3. Abundant in some barren fields near Raglan, 1850, *Woods.*

Valerianella dentata (L.) Poll.   Tooth-fruited Corn-salad

Cultivated ground; rare. Colonist.
1. Tregare*; Duffield's Farm, Upper Redbrook*, *Charles.* 4. Near Tintern, *Watkins:* Mounton*; Llanvair Discoed; Llanmelin*; near St. Lawrence, Chepstow, *Shoolbred.*

var. **mixta** (Vahl) Dufr.
4. By Thicket Wood, Rogiet*.

VALERIANA L.

Valeriana officinalis L.   Great Valerian

Damp woods, ditches and reen-sides; common. All districts. Native.

Valeriana dioica L.   Marsh Valerian

Marshes, damp woods and stream-sides; locally frequent. Native.
1. Abergavenny district, *White:* Honddu and Grwyne Fawr Valleys, *Ley:* Wye Valley, near Monmouth, *Hamilton:* Beaulieu Farm, Monmouth*; by the River Monnow, Osbaston*, *Charles:* near Triley Bridge*; Pont-y-spig. 2, Craig-yr-Hafod, near Blaenavon, *Guile.* 3. Near Usk, *Clark:* near Ynys-y-fro Reservoir, *Hamilton:* Raglan district, *Miss Frederick,!:* Clearwell, near Michaelstone y Vedw; by Coed-y-llyn, Llanover*; canal side Goytre*; Llandegfedd; near Pentwyn Farm, Llanddewi Fach; near Castell-y-bwch*. 4. Near Fair Oak Pond, Tintern*; Chepstow Park Wood*; Shirenewton*; The Glynn; Rhyd-y-fedw; Kilgwrrwg; Dinham, *Shoolbred:* Trelleck Bog, *Charles:* Llwynycelyn, *T. G. Evans.*

CENTRANTHUS DC.

Centranthus ruber (L.) DC.   Red Valerian

Walls, railway banks and cuttings, and limestone cliffs; frequent in District 4; rarer elsewhere. Denizen.
1. Wyesham, *Charles.* 2. Abercarn. 3. Near Newport, *Hamilton:* Usk, *Miss E. A. Jenkins:* Plas Machen; Llanbadoc. 4. Chepstow*, *Conway, Winch, Hamilton, Shoolbred,!:* Tintern Parva, *Hamilton:* Tintern, *Shoolbred:* Mathern, *Godbey:* Llanvair Discoed; Penhow*; Magor.

DIPSACACEAE
DIPSACUS L.

Dipsacus fullonum L. ssp. **fullonum**   Wild Teasel
D. sylvestris Huds.

Ditches, reens, wood borders, river banks and field borders; common in District 5; locally frequent in Districts 1 and 4; rare in Districts 2 and 3. Native.

Dipsacus pilosus L.   Small Teasel

Woods, river banks, railway banks and quarries; locally frequent in and almost confined to the Wye Valley. Native.
1. Banks of the River Wye, near Monmouth*, *Bryan:* Monmouth, *Hamilton:* between Monmouth and the Bibblings; Hadnock; Lady Park Wood, *Charles,!:* near Pont-y-spig*. **3.** Usk, *Forster, Mrs. Ellis:* Llancayo, near Usk, *Clark.* **4.** Near Chepstow, *Forster, Shoolbred:* near Tintern Abbey, *Poole:* near Gwernesney†, *Clark:* between Wynd Cliff and Tintern; Barbadoes Hill; near Shirenewton; Itton; near Penterry; Mounton Valley*, *Shoolbred:* Mitchel Troy, *Charles:* by Cliff Wood, Mounton*.

KNAUTIA L.

Knautia arvensis (L.) Coult.   Field Scabious
Scabiosa arvensis L.

Hedgebanks, pastures, meadows, railway banks and wood borders; common. Districts 1-4. Native.

SCABIOSA L.

Scabiosa columbaria L.   Small Scabious

Grassy limestone banks; rare. Native.
4. Wynd Cliff*, *Conway, Shoolbred:* Tintern Abbey, *Clark:* Chepstow, *Clark, Shoolbred.*

Hamilton's records from Liswerry, Llanwern and Caerleon require confirmation; the first is almost certainly erroneous.

SUCCISA Haller

Succisa pratensis Moench   Devil's-bit Scabious
Scabiosa succisa L.

Moist meadows and pastures, marshy places, open woods and moorland; common. All districts. Native.

COMPOSITAE
BIDENS L.

Bidens cernua L.   Nodding Bur-Marigold

Marshes, pond margins and by reens and canals; rare. Native.

172

COMPOSITAE

1. Monmouth, *Charles:* Llanfoist. **3.** Llantarnam, *Reader:* Malpas, *Hamilton,!.* **5.** St. Brides Wentlloog, *Storrie, Hamilton:* Marshfield*, *Hamilton,!:* Whitewall Common, Magor*, *R. Lewis.*

Bidens tripartita L.　Three-cleft Bur-Marigold
Reen sides, marshes, riversides and margins of ponds; widespread but not common. Native.
1. Monmouth*; Hadnock Stream, near Monmouth*; between Monmouth and Redbrook*, *Charles:* between Mitchel Troy and Dingestow*, *R. Lewis:* Llanfoist; Dingestow. **2.** Canal, Newport to Abercarn, *McKenzie:* Risca*. **3.** Llanbadoc; Usk, *Clark:* Llantarnam; Malpas, *Hamilton:* Raglan district, *Miss Frederick.* **4.** Llanllywel, *Clark:* Mathern*, *Shoolbred:* Wentwood, *Hamilton:* St. Brides Netherwent, *Evans:* Whitebrook*; Penallt*, *Charles.* **5.** Rumney*.

Bidens frondosa L.
Alien.
5. Newport Docks, 1954, *McClintock.*

AMBROSIA L.

Ambrosia artemisiifolia L.
Alien.
5. Alexandra Dock, Newport*, 1942, *Macqueen.*

SENECIO L.

Senecio jacobaea L.　Ragwort
Pastures, open woods, railway banks and waste places; common. All districts. Native.

Senecio aquaticus Hill　Marsh Ragwort
Marshes, margins of ponds, rivers and streams; common. All districts. Native.

Senecio erucifolius L.　Hoary Ragwort
Roadsides, banks, woods, quarries and railway banks; locally frequent to locally common, especially on the limestone. Native.
1. Fairly common. **2.** Machen. **3.** Pontypool Road; Cwrt Henllys. **4.** Common. **5.** Common.

Senecio squalidus L.　Oxford Ragwort
Roadsides, railway banks and tracks, waste ground and walls; common. Established alien.

173

COMPOSITAE

First recorded by Dr. J. S. Clarke in 1904 from near the Severn Tunnel Junction. This species has spread over the county within the past 50 years and has been reported from Denny Island.

Senecio sylvaticus L.   Heath Groundsel

Wood borders, heaths and dry banks; locally frequent. Native.
1. Beaulieu Wood; Redding's Enclosure; Garth Wood, near Monmouth; Hadnock Farm; Hadnock Wood*, *Charles*. **2.** Blaenavon, *Clark:* Cwmcarn, *Miss Jones:* Mynydd Dimlaith; near Graig Gwyn, Abercarn*. **3.** Llanfrechfa Lower*. **4.** Wentwood, *Clark:* above Llandogo, . *Ley:* between Rogiet and Llanvair Discoed, *Dr. Clarke:* Wynd Cliff; between Tintern and Trelleck*; Rogiet*, *Shoolbred:* Hael Wood, Penallt*; Trelleck Beacon,* *Charles*.

Senecio viscosus L.   Sticky Groundsel

Railway tracks, colliery tips, roadsides and wood clearings; locally frequent. Doubtful Native.
1. Hadnock Siding*; Hadnock Wood*; Troy Station, Monmouth*, *Charles:* Abergavenny, *Hardaker,!*. **2.** Frequent to common on old colliery tips. **3.** Usk, *Charles*. **4.** Between Bigsweir and Redbrook, *Ley:* Severn Tunnel Junction*; Tintern, *Shoolbred:* near Redbrook*; Mitchel Troy*, *Charles*. **5.** Newport Docks, *Hamilton:* between Portskewett and Caldicot Pill, *Shoolbred:* Rumney.

Senecio vulgaris L.   Groundsel

Cultivated and waste ground, and roadsides; very common. All districts. Native.

var. **radiatus** Koch.
Frequent to locally common.

DORONICUM L.

Doronicum pardalianches L.   Great Leopard's-bane

Hedgebanks, railway banks, wood borders and waste ground; rare. Denizen.
1. Beaulieu Wood*; Kymin Hill*; Hadnock*; *Charles*. **4.** Newchurch East*, *Shoolbred:* Upper Redbrook*; Troy Park Wood, *Charles:* Itton*, *Shoolbred, T. G. Evans*.

TUSSILAGO L.

Tussilago farfara L.   Coltsfoot

Waste and cultivated ground, railway banks and roadsides; very common. All districts. Native.

174

COMPOSITAE

PETASITES Mill.

Petasites hybridus (L.) Gaertn., Mey. & Scherb. Butterbur
River banks and borders of wet fields; locally frequent. Native.
1. Llanthony, *Ley:* Monmouth*; *E. Lees, Charles:* Skenfrith*; Osbaston*;
Hadnock*, *Charles:* near Govilon, *Mrs. Scholberg:* Abergavenny. **3.** By
the River Rhymney, near the entrance to 'The Gaer', *Hamilton:* Usk,
Smith,!: Llanbadoc, *Briggs,!:* near Pontnewydd; by the Afon Lwyd,
Llanfihangel Llantarnam. **4.** Tintern; Penallt, *Charles:* St. Brides Nether-
went, *Mrs. Ellis:* near Bigsweir Station, *T. G. Evans:* Magor.

Petasites fragrans (Vill.) C. Presl   Winter Heliotrope
Roadsides and hedgebanks; frequent. Denizen.
1. Leasebrook Lane, Dixton*; Coleford Road, near Monmouth*,
Charles. **3.** Bassaleg; Allt-yr-yn, *McKenzie,!:* Helmaen, *Charles:* Usk
district, *Mrs. Hall:* near Castleton; between Coed Mawr and Laswern
Wood, Goytre. **4.** About Chepstow, *Shoolbred, T. G. Evans:* Llandogo*,
Charles: between Langstone and Penhow; St. Brides Netherwent. **5.**
Goldcliff.

INULA L.

Inula helenium L.   Elecampane
Pastures, roadsides and woods; rare to locally frequent. Denizen.
1. On the road to Skenfrith, beyond Llanvetherine, *E. Lees:* Abergavenny
district, *White:* Llanvihangel-Ystern-Llewern, *Ley:* between Skenfrith
and Grosmont, *C. Parkinson:* Skenfrith*, *Perman:* Tregare, *Miss Frederick:*
The Hendre*; near Norton*; Wonastow*, *Charles:* **4.** Llanvair Discoed,
Forster: Llangwm, *Clark:* St. Arvans; *Hamilton, Shoolbred:* Dinham,
Shoolbred: Pen-y-clawdd*, *Jacques, Charles:* Wonastow, *Charles:* Run
stone, *Shoolbred, T. G. Evans:* Bishton; near Llanwern.

Inula conyza DC.   Ploughman's Spikenard
Inula squarrosa (L.) Bernh.
Roadside banks, quarries, railway banks and woods; locally frequent and
almost restricted to limestone soils. Native.
1. Monmouth; Hadnock Quarry*; Upper Redbrook*; near Nant-y-
gern, Newcastle*; Dixton*; near Abergavenny Junction*, *Charles.*
2. Pontypool*, *T. H. Thomas:* Crosskeys.* **4.** Wynd Cliff*, *Bailey,*
Shoolbred: near Chepstow; Pen-moel Quarries, *Hamilton:* Shirenewton;
Mounton; Park Wall Hill; St. Pierre; near Runstone, *Shoolbred:* Llan-
melin*, *Miss Cooke:* between Whitebrook and Redbrook*; Penallt*;
near Troy House, *Charles:* Rogiet, *Mrs. Ellis:* Ifton*; Slade Wood;
Penallt Common.

175

COMPOSITAE

PULICARIA Gaertn.

Pulicaria dysenterica (L.) Bernh. Fleabane

Wet roadsides and fields, and river banks; common, especially on the marine alluvium of District 5. All districts. Native.

Pulicaria vulgaris Gaertn. Small Fleabane

Casual.

5. Alexandra Dock, Newport, *Hamilton.*

FILAGO L.

Filago vulgaris Lam. Upright Cudweed
Filago germanica L.

Cultivated ground, quarries and roadsides; rare to locally frequent. Native.

1. Garth Wood; Lady Park Wood; New Hill, Grosmont*; Cwmyoy*; quarry under Craig Syfyrddin*, *Charles.* **3.** Bettws, *Hamilton:* **4.** About Chepstow, *Shoolbred:* Llanwern, *Hamilton.*

Filago minima (Sm.) Pers. Small Cudweed

Old colliery tips, railway tracks and gravelly places; rare. Native.
1. Abergavenny district, *White.* **2.** Crosskeys*; near Bedwellty Pits, near Tredegar; below Mynydd Dimlaith*.

GNAPHALIUM L.

Gnaphalium sylvaticum L. Heath Cudweed

Heaths and rough pastures; rare to locally frequent. Native.
1. Cwm Bwchel, Llanthony, *Ley:* Kymin Hill, Monmouth*; Half-way House Wood, Monmouth*; Beaulieu Farm, Monmouth*; Redding's Enclosure; New Hill, Grosmont*; Hadnock Wood*; Lady Park Wood; Pritchard's Hill*, *Charles.* **3.** Between Estavarney Farm and Goytre, *Clark:* Foxwood, *Hamilton.* **4.** Beacon Hill, Trelleck, *Ley:* Wentwood*, *Shoolbred, Hamilton:* Bulmore, *Hamilton:* The Barnetts Farm, near Chepstow; Barbadoes Hill, Tintern; near Trelleck Bog; Devauden; near Pandy Mill, Itton*; Chepstow Park Wood*; Ifton, *Shoolbred:* Llanvair Discoed, *Lannon:* near Caer Lieyn, Kemeys Inferior.

Gnaphalium uliginosum L. Cudweed

Cultivated ground, muddy field paths and damp roadsides; common in all districts. Native.

COMPOSITAE

ANAPHALIS DC.

Anaphalis margaritacea (L.) Benth. American Cudweed

Colliery tips, railway banks, river banks, open woods and waste ground; locally common. Denizen.
1. Beaulieu Wood*; near Highmeadow Siding; Redding's Enclosure*; near Hadnock; near Trebella, *Charles:* near The Slaughter, within Monmouthshire, *Trapnell.* **2.** Common, especially in the Rhymney Valley. **3.** Near Michaelstone y Vedw, *Babington:* between Caerleon and Newport, *Clark:* Raglan district, *Miss Frederick:* Wern Fawr, near Nant-y-derry, *Briggs.* **4.** Trelleck Beacon*, *Charles:* between Moss Cottage and St. Arvans*, *R. Lewis:* near Pen-yr-heol, *Mrs. Parris.* **5.** Alexandra Dock, Newport*, *McQueen:* Spencer Steel Works site*.

First recorded in Ray, *Synopsis*, ed. 3, 182, 1724 from the banks of the River Rhymney, on the authority of Edward Lhwyd.

ANTENNARIA Gaertn.

Antennaria dioica (L.) Gaertn.   Mountain Everlasting

Mountain heaths on limestone; very rare. Native.
2. About Abersychan, *Clark:* Blorenge; Twm Barlwm, Risca, *Hamilton.*

SOLIDAGO L.

Solidago virgaurea L.   Goldenrod

Woods, river banks, hedgebanks and heaths; frequent to locally common. Districts 1-4.   Native.

Solidago canadensis L.   Canadian Goldenrod

Alien.
5. Waste ground about Newport*.

ASTER L.

Aster tripolium L.   Sea Aster

Salt marshes and tidal banks of rivers; locally common, extends up the River Wye to above Tintern and up the River Usk to Caerleon. Districts 3-5. Native.
The variety without ray florets (var. *discoideus* Reichb.) is not uncommon.

Aster novi-belgii L.

Naturalized garden outcast.
4. Quarry, Ifton*. **5.** Spencer Steel Works site*.

177

COMPOSITAE

ERIGERON L.

Erigeron acer L.   Blue Fleabane

Roadside banks, railway banks and quarries on the limestone, and walls; rare. Native.

1. Railway bank opposite The Bibblings*; Lady Park Wood*; Redding's Enclosure; between Dingestow and Mitchel Troy*, *Charles*. **2.** Little Mountain, Trevethin*, *T. H. Thomas*. **3.** Trostrey, *Clark*. **5.** Near Caldicot Tin Plate Works*, *Shoolbred*.

CONYZA Less.

Conyza canadensis (L.) Cronq.   Canadian Fleabane
  Erigeron canadensis L.

Waste ground; locally frequent. Alien.

1. Abergavenny. **3.** Canal path, Malpas Road, Newport. **4.** Redbrook*, *Charles:* Tintern, *T. G. Evans:* Chepstow*. **5.** Alexandra Dock, Newport*, *Macqueen:* Spencer Steel Works site*.

BELLIS L.

Bellis perennis L.   Daisy

Pastures and grassy places; very common in all districts. Native.

EUPATORIUM L.

Eupatorium cannabinum L.   Hemp Agrimony

Wet woods, streamsides and marshy places; common in Districts 1-4, frequent in District 5. Native.

ANTHEMIS L.

Anthemis tinctoria L.

Waste ground. Casual.
5. Newport Docks*, *Smith*.

Anthemis cotula L.   Stinking Mayweed

Cultivated ground and roadsides; frequent to common. All districts. Colonist.

Anthemis arvensis L.   Corn Chamomile

Cultivated ground; rare. Colonist.
4. The Lindors Farm, near Chepstow, *Shoolbred*.

COMPOSITAE

CHAMAEMELUM Mill.

Chamaemelum nobile (L.) All.   Chamomile
Anthemis nobilis L.

Grassy places near the sea; very rare. Native.
Included in a list of plants observed in the neighbourhood of Pontnewydd by J. H. Clark, published in Watson's *New Botanist's Guide*, 1837. Since the list contains other maritime plants the record may be correct.

ACHILLEA L.

Achillea millefolium L.   Yarrow

Pastures, meadows, grassy roadsides and wall tops; very common. All districts. Native.

Achillea ptarmica L.   Sneeze-wort

Marshy pastures and roadsides, and river banks; frequent to locally common. Native.
1. Abergavenny, *White:* banks of the Wye, Monmouth; near Hadnock Stream; Dixton; Coed-anghred Hill, Skenfrith*, *Charles.* **2.** Rhymney Bridge*; Blean-y-Cwm, Dukestown*; near Tredegar*. **3.** Glascoed, *Clark:* Llangattock; near Castell-y-Bwch; Christchurch; Llandegfedd; Llanddewi Fach*. **4.** Fairly common in the Chepstow district; Bigsweir, *Shoolbred:* near Llandogo, *Mrs. Leney:* The Narth, *T. G. Evans.* **5.** Rumney*; near Marshfield*.

Achillea ligustica All.

Waste ground. Alien.
5. Newport Docks*, *Macqueen.*

TRIPLEUROSPERMUM Schultz Bip.

Tripleurospermum maritimum (L.) Koch ssp. **maritimum**
Matricaria maritima L.   Scentless Mayweed

Banks of the Severn and sea walls; frequent. Native.

ssp. **inodorum** (L.) Hyland. ex Vaarama
Matricaria inodora L.
Waste and cultivated ground, roadsides and railway banks; common in all districts. Native.

MATRICARIA L.

Matricaria recutita L.   Wild Chamomile
M. chamomilla auct.

Cultivated ground and roadsides; common in Districts 4 and 5, frequent in Districts 1 and 3, unrecorded from District 2. Colonist.

COMPOSITAE

Matricaria matricarioides (Less.) Porter   Rayless Chamomile
Matricaria discoidea DC.

Roadsides, cart tracks, waste and cultivated ground; common in all districts. Colonist.

First recorded by Miss E. Armitage in 1901. Introduced from North America towards the end of the nineteenth century.

CHRYSANTHEMUM L.

Chrysanthemum segetum L.   Corn Marigold

Cultivated ground; locally common. Colonist.
1. Abergavenny, *Hamilton:* Llanfoist; near Coed y Gatlas, Llangattock*.
3. Near Caerleon. **4.** Portskewett, *Hamilton:* Caldicot; Ifton Hill, *Shoolbred:* Caerwent*, *Benison:* Trelleck*, *Charles:* near Upper Grange, Llanfihangel; between St. Brides Netherwent and Thicket Wood*.

LEUCANTHEMUM Mill.

Leucanthemum vulgare Lam.   Dog Daisy, Ox-eye Daisy
Chrysanthemum leucanthemum L.

Pastures, meadows, railway banks, grassy roadsides and walls; common. All districts. Native.

TANACETUM L.

Tanacetum parthenium (L.) Schultz Bip.   Feverfew
Chrysanthemum parthenium (L.) Bernh.

Roadsides, hedgebanks, walls and waste ground; frequent, especially near habitations. All districts. Alien.

Tanacetum vulgare L.   Tansy

River banks, railway banks, roadsides and waste ground; common by rivers in Districts 1-4, rare in District 5. Native.
5. Pen-carn, near Coedkernew, *Hamilton:* Newport Docks.

ARTEMISIA L.

Artemisia vulgaris L.   Mugwort

Waste ground, roadsides and river banks; common. All districts. Native.

Artemisia absinthium L.   Wormwood

River banks, railway banks, roadsides and colliery tips; rare to locally frequent. Denizen.
1. Llanthony Abbey, *Ley.* **2.** Machen, *Hamilton:* west side of the Blorenge, *Charles:* below Mynydd Dimlaith, *Mrs. Pinkard:* near Pandy Mawr,

COMPOSITAE

Bedwas; between Brynmawr and Beaufort*; Rhymney*; Dukestown*; Crosskeys. **3.** Pontnewydd*, *Conway:* River Ebbw, Newport; Malpas; Castleton, *Hamilton:* St. Mellons; Llanrumney; between Castleton and St. Mellons. **4.** Near Caldicot Castle, *Warner:* Trelleck, *Clark:* Redbrook*, *Charles:* Magor, *E. Nelmes.* **5.** Near Newport, *Hamilton:* Newport Docks, *Davies:* Peterstone Wentlloog, *Mrs. Adams.*

First recorded by Richard Warner in 1797.

Artemisia maritima L.   Sea Wormwood

Tidal river banks and salt marshes; locally common. Native.
4. About Chepstow, *E. Lees:* by the River Wye, between Chepstow and Piercefield. **5.** Mouth of the River Ebbw, *Hamilton:* near Magor*; Rogiet reens; St. Pierre Pill, *Shoolbred:* Mathern*, *Shoolbred, T. G. Evans:* Magor Pill, *T. G. Evans:* Peterstone Wentlloog; Collister Pill, near Undy.

CARLINA L.

Carlina vulgaris L.   Carline Thistle

Pastures, dry banks and railway banks on calcareous soils, sea walls and limestone quarries; rare to locally frequent. Native.
1. Near Llanthony, *Ley:* opposite the Bibblings; near Pentre Farm, The Hendre*, *Charles:* Seargent's Grove, near Kings Wood. **2.** Between Crosskeys and Wattsville, *McKenzie:* Trefil; Machen*. **3.** Rhadyr, near Usk, *Clark:* Raglan district, *Miss Ferderick.* **4.** Wynd Cliff, *Watkins, Hamilton, Shoolbred,!:* by Halfway House, Redbrook, *Charles:* Pen-y-clawdd, *Miss Davis:* Highmoor Hill, *Mrs. Ellis:* Llanwern. **5.** Sea wall, Rumney*.

ARCTIUM L.

Arctium lappa L.   Greater Burdock
A. majus Bernh.

Wood borders, hedgebanks, rough pastures and reen banks; rare to locally frequent. Native.
1. Near Wyesham; Llantilio Crossenny. **3.** Cilfeigan*; The Forest, Llangybi*. **4.** Mathern; Tintern; near Pandy Mill, Itton*; Slade Wood, Rogiet; Usk Road, near Chepstow, *Shoolbred:* Troy House, *Riddelsdell:* Llanwern*. **5.** Near Rumney; near Goldcliff; near Whitson; Undy; near Marshfield.

Arctium minus Bernh.   Lesser Burdock

Roadsides, hedgebanks, woods and reen banks. Native.

ssp. **minus**

Frequent to locally common. All districts.

COMPOSITAE

ssp. **nemorosum** (Lej.) Syme
1. Near Llangattock Vibon Avel; near The Queen's Head, Honddu Valley, *Ley:* Llanfihangel Crucorney, *Miss Dawber.* **5.** Near Llanwern*.

CARDUUS L.

Carduus tenuiflorus Curt.   Slender-headed Thistle

Sea walls, river banks and waste ground; rare. Native.
5. Newport*, *Conway, Clark:* near the mouth of the River Ebbw, *Hamilton:* Marshfield; Peterstone Wentlloog, *Richards,!:* Denny Island, *Matthews:* Rumney*.

Carduus nutans L.   Musk Thistle

Heaths, pastures, roadsides and quarries, especially on calcareous soils; rare to locally frequent. Native.
1. Abergavenny district, *White:* Hadnock*; Duffield's Farm, near Monmouth*; Chippenham*, *Charles.* **2.** Risca*; Rhymney*. **3.** Caerleon; Llanfrechfa; Mendalgief, *Hamilton:* Raglan district, *Miss Frederick.* **4.** Portskewett; Usk Road, near Chepstow; Howick, *Shoolbred:* Ifton*. **5.** Newport, *Clark:* Liswerry, *Hamilton.*

Carduus acanthoides L.   Welted Thistle
   C. crispus auct.

Roadsides, waste ground and rough pastures; frequent. All districts. Native.

CIRSIUM Mill.

Cirsium eriophorum (L.) Scop.   Woolly-headed Thistle

Rough pastures on calcareous soils; rare, Native.
1. Cefn-garw, near Tregare, *Clark.* **4.** Itton*, *Barthrop, Shoolbred:* near Howick; near Pandy Mill, *Shoolbred:* Wonastow, *Charles:* near Severn Tunnel Junction.

Cirsium vulgare (Savi) Ten.   Spear Thistle
   C. lanceolatum (L.) Scop. non Hill

Pastures, roadsides and waste ground; very common. All districts. Native.

Cirsium palustre (L.) Scop.   Marsh Thistle

Wet pastures, marshes and streamsides; very common. All districts. Native.

Cirsium arvense (L.) Scop.   Field Thistle

Pastures, roadsides and cultivated ground; very common. All districts. Native.

var. **mite** Koch
4. Below Penmoel*, *Shoolbred.*

var. **setosum** C. A. Mey.
3. Pontypool*, *Sister Albertine.*

Cirsium acaule Scop.   Stemless Thistle
Heaths, dry pastures, banks and quarries on calcareous soils; locally common in District 4, rare elsewhere. Native.
1. War Fields, Staunton Road, near Monmouth*; Wyesham*, *Charles.*
2. Trefil*. **4.** Near Chain Bridge, Kemeys, *Clark:* Itton, *Hamilton:* Piercefield; Wynd Cliff; The Barnetts; Chepstow*; Dinham; Little Dinham, *Shoolbred:* The Dingle, Llangoven*, *Miss Heeps:* near Rogiet, *Mrs. Ellis.*

var. **caulescens** Pers.
4. Beteeen St. Arvans and Chepstow Park*, *Shoolbred.*

Cirsium dissectum (L.) Hill   Meadow Thistle
C. pratense (Huds.) Druce non DC.
Marshes, bogs and wet meadows; locally frequent. Native.
1. Grwyne Fawr Valley; near Pont-y-spig, *Ley,!*. **2.** Near Pentwyn-mawr, Abercarn, *Mrs. Leney:* Trefil; Pen-y-fan Pond. **3.** Near Pontnewydd, *Conway:* Ynys-y-fro; Foxwood; Henllys, *Hamilton:* Raglan district, *Miss Frederick:* near Llantarnam*; north of Mescoed Mawr; Garw Wood, Pontnewydd*; near Castell-y-bwch*. **4.** Pen-y-cae-mawr, near Wentwood, *Clark:* near Rhyd-y-fedw*; below Chepstow Park; between Magor and Undy*; near Chepstow Park Wood*; The Glyn, Itton*; Kilgwrrwg, *Shoolbred:* Trelleck Bog, *Shoolbred,!.*

SILYBUM Adans.
Silybum marianum (L.) Gaertn.   Milk Thistle
Waste ground; rare. Alien.
1. Abergavenny, *White:* Chippenham*, *Charles.* **4.** Garden weed, Chepstow*, *Shoolbred.* **5.** Newport, *Clark.*

ONOPORDUM L.
Onopordum acanthium L.   Scotch Thistle
Waste ground and dock ballast; rare. Alien.
1. Monmouth, *Hamilton.* **2.** Pontllan-fraith, *Miss Thomas.* **3.** Usk, *Fraser.* **4.** Below Newchurch, *Fraser.* **5.** Newport, *Clark, Hamilton:* near Caldicot*, *Miss E. David.*

CENTAUREA L.
Centaurea scabiosa L.   Greater Knapweed
Grassy roadsides, railway banks, dry pastures and quarries; locally common, especially on calcareous soils. Native.

COMPOSITAE

1. Abergavenny, *Miss Frederick,!:* Monmouth*, *Miss Frederick, Charles.*
2. Machen, *Hamilton:* near Trehir Quarry, Bedwas*; Risca. 3. Michaelstone y Vedw; Bassaleg; St. Woolas, *Hamilton.* 4. Between Chepstow and Itton, *Clark:* fairly common about Chepstow, *Shoolbred:* near Carrow Hill. 5. Dyffryn, *Hamilton.*

Centaurea montana L.
Garden escape.
3. Bank of the River Rhymney, Llanrumney*.

Centaurea cyanus L. Cornflower
Waste and cultivated ground; rare. Alien or colonist.
3. Cornfield, Penylan Farm, Llandegfedd, *Campbell.* 4. Near Magor†, *Clark:* Caerwent; Dewstow; Portskewett, *Hamilton:* Dinham, *Shoolbred:* Upper Redbrook*, *Charles.*

Centaurea nigra L. Black Knapweed
Pastures, grassy roadsides and hedgebanks; common. All districts. Native.

Centaurea nemoralis Jord.
Pastures, grassy roadsides and open woods; locally common. Districts 1, 3-5. Chiefly on calcareous soils. Native.

Centaurea calcitrapa L.
Alien.
5. Newport Docks, *Clark.*

SERRATULA L.

Serratula tinctoria L. Saw-wort
Open woods, hedgebanks and pastures; locally common to locally frequent, chiefly on calcareous soils. Native.
1. Lady Park Wood, *Charles.* 2. Rhymney Valley, *Hamilton:* near Pentwynmawr, Abercarn, *Mrs. Leney.* 3. Usk, *Hamilton:* Raglan district, *Miss Frederick:* Garw Wood, near Pontnewydd; Llandegfedd*; Llanddewi Fach. 4. Near Chepstow*, *Clark, Hamilton, Shoolbred, Elliot:* Barnett Woods, *Shoolbred:* plentiful in some places in the Chepstow district, *Shoolbred:* Rogiet, *Mrs. Ellis:* Mounton, *T. G. Evans.*

CICHORIUM L.

Cichorium intybus L. Chicory
Rough pastures, waste and cultivated ground; rare. Alien.
1. Chippenham*; near Fiddler's Elbow, Monmouth*, *Charles.* 4. Shirenewton, *Hamilton:* The Barnetts Farm, near Chepstow*; Portskewett; Dinham; near Moulton, *Shoolbred:* near Caldicot Castle, *Mrs. Ellis.* 5. Newport Docks, *McKenzie.*

COMPOSITAE

LAPSANA L.

Lapsana communis L.   Nipplewort

Hedgebanks, roadsides and waste places; very common. All districts. Native.

HYPOCHOERIS L.

Hypochoeris radicata L.   Long-rooted Cat's-ear

Pastures, meadows and grassy roadsides; very common. All districts. Native.

LEONTODON L.

Leontodon autumnalis L.   Autumnal Hawkbit

Pastures, meadows, grassy roadsides and colliery tips; common. All districts. Native.

Leontodon hispidus L.   Hairy Hawkbit

Pastures, meadows, grassy roadsides, quarries and hedgebanks; common. All districts. Native.

Leontodon taraxacoides (Vill.) Mérat   Hairy-headed Hawkbit
   L. leysseri G. Beck

Pastures, grassy banks and heaths; frequent. All districts. Native.

PICRIS L.

Picris echioides L.   Ox-tongue

Roadsides, banks, cultivated ground, waste places and quarries; locally frequent on calcareous soils and marine alluvium. Native.
1. Near Fiddler's Elbow, Monmouth, *Charles*. **3.** Raglan district, *Miss Frederick*. **4.** Chepstow*, *Clark, Shoolbred:* Tintern, *Hamilton:* Magor*, *Shoolbred:* Severn Tunnel Junction, *Mrs. Ellis:* near Llanwern; Piercefield Wood. **5.** Pillgwenlly, Newport*, *Conway:* Newport, *Clark, Hamilton, Smith,!:* between Pont Ebbw and the Lighthouse, *McKenzie:* Marshfield*; near Whitson*; Goldcliff; by Greenland Reen, St. Mellons*; Traston, near Newport.

Picris hieracioides L.   Hawkweed Ox-tongue

Roadsides, hedgebanks, wood borders and railway banks; locally common, especially on calcareous soils. Native.
1. Monmouth, *Ley, Charles:* near Hadnock*; near Pentre Farm, Hendre*; Wyesham*; Osbaston*; Grosmont Road, near Norton*; Llanvapley*; White Castle*; between the Onen and Penrhos*; Rockfield*; Newcastle*; Llantilio Crossenny, *Charles:* Tregare; Lady Park Wood*. **3.** Abergavenny

185

COMPOSITAE

Road, near Usk, *Clark:* St. Julians; Michaelstone y Vedw, *Hamilton:* Raglan district, *Miss Frederick:* near Cilfeigan Park*; near Coed-y-paen. **4.** Langstone, *Hamilton:* Chepstow; Wynd Cliff*; Mounton; Rogiet; The Coombe; Langham's Lane, *Shoolbred:* near Whitebrook*; Mitchel Troy*; between Trelleck and Cleddon*, *Charles:* near Tymawr, Penallt, *Beckerlegge.* **5.** Spencer Steel Works site*.

TRAGOPOGON L.

Tragopogon pratensis L. Goatsbeard

Pastures, roadsides and waste ground. Native.

ssp. **pratensis**
Rare.
3. Pontypool Road Station*. **4.** Portskewett, *Shoolbred.*

ssp. **minor** (Mill.) Wahlenb.
Frequent to locally common. All districts.

Tragopogon porrifolius L. Salsify

Marshy meadows and sea banks; rare. Denizen.
3. Near Malpas*. **4.** By the Wye below Tintern Abbey*, probably an escape from past cultivation at the Abbey, *Babington, Southall, Shoolbred*: near Chepstow*, *Shoolbred.* **5.** Rumney*, *Miss Ogden.*

Tragopogon crocifolius L.

Alien.
4. Grew in abundance up to 1938 at the Bulwark, Chepstow, recorded by Capt. Fraser. The site is now built over.

LACTUCA L.

Lactuca serriola L. Prickly Lettuce

Limestone quarries, waste ground and railway ballast; rare. Alien.
4. Near Severn Tunnel Junction; Rogiet*, *Shoolbred:* Ifton*. **5.** Undy; Traston, Newport; Spencer Steel Works site*.

(**Lactuca virosa** L. Recorded by Shoolbred from near the Severn Tunnel Junction, but the specimen in his herbarium is *L. serriola.*)

MYCELIS Cass.

Mycelis muralis (L.) Dumort. Wall Lettuce
  Lactuca muralis (L.) Gaertn.

Walls and rocky places; common. All districts. Native.

COMPOSITAE

SONCHUS L.

Sonchus arvensis L.   Corn Sowthistle

Cultivated ground and roadsides; frequent to locally common. Districts 1, 3-5. Native.

Sonchus oleraceus L.   Common Sowthistle

Cultivated and waste ground; very common. All districts. Native.

Sonchus asper (L.) Hill   Rough Sowthistle

Cultivated and waste ground; common. All districts. Native.

HIERACIUM L.

Hieracium lasiophyllum Koch.

Limestone banks; rare. Native.
1. Monmouth, *Ley.*

Hieracium schmidtii Tausch

Rocky limestone woods; rare. Native.
4. Wynd Cliff, *Shoolbred:* by Black Cliff Wood, near Tintern, *Redgrove.*

Hieracium pachyphylloides (Zahn) Roffey
H. pachyphyllum (Purchas) F. N. Williams non Brenner

Rocky limestone woods; rare. Native.
4. Piercefield, *Shoolbred.*

Hieracium pellucidum Laest.

Limestone and sandstone cliffs, rocks and walls; locally frequent. Native.
P. Llanthony Abbey, *Bicheno, teste Ley.* **4.** Near Old Furnace, Tintern*;
1ont y Saison; Llandogo, *Shoolbred:* Tintern Road, Chepstow, *Redgrove.*

Hieracium asteridiophyllum Sell & West
H. pellucidum var. *lucidulum* W. R. Linton *pro parte*

Mountain cliffs; rare. Native.
1. Tarren yr Esgob, Honddu Valley, *Ley.*

Hieracium exotericum Jord. ex Bor.

Rocky places; frequent. Native.
1. Abergavenny, *Druce:* near Great Goytre*; Tarren yr Esgob*; near Llanthony*; Grosmont*. **2.** Machen, *Riddelsdell.* **3.** Between Pen-y-lan and Pennsylvania, near Bassaleg*; Llanrumney*. **4.** Wynd Cliff, *T. G. Evans.*

COMPOSITAE

Hieracium grandidens Dahlst.

Roadside banks; locally frequent. Native.
3. Near Penrheol-y-badd-fawr, Henllys*. **4.** Between Trelleck and Llan-dogo; Tintern Road, near Black Cliff, *Shoolbred, Miss M. Cobbe:* Wynd Cliff, *Riddelsdell:* Wentwood*.

Hieracium sublepistoides (Zahn) Druce

Walls, rocky places and banks; frequent. Districts 1-4. Native.

Hieracium glevense (Pugsl.) Sell & West

Roadside banks; rare. Native.
4. Between Coad Ithel and Llandogo, *Shoolbred.*

Hieracium cuneifrons (W. R. Linton) Pugsl.

Rare. Native.
4. Near Moss Cottage, Wynd Cliff, Llandogo, *Shoolbred.*

Hieracium euprepes F. J. Hanb.

Old Red Sandstone cliffs; rare. Native.
1. Black Mountains, Honddu Valley, *Ley.*

Hieracium cacuminum (A. Ley) A. Ley

Limestone banks; rare. Native.
4. Wynd Cliff, *Shoolbred.*

Hieracium leyanum (Zahn) Roffey

Old Red Sandstone cliffs; rare. Native.
1. Tarren yr Esgob, Honddu Valley, *Ley.*

Hieracium surrejanum F. J. Hanb.

Limestone cliffs; rare. Native.
4. Wynd Cliff, *Pugsley.*

Hieracium subamplifolium (Zahn) Roffey

Limestone banks and cliffs; locally frequent. Native.
4. Near Tintern, *Bickham & Ley:* Wynd Cliff, *Shoolbred, Marshall:* between Pen-y-parc and Devauden, *Shoolbred.*

Hieracium submutabile (Zahn) Pugsl.

Walls; local. Native.
4. Tintern, *Townsend.*

Hieracium diaphanoides Lindeb.

Banks, quarries and waste ground; locally frequent. Native.
2. Pandy Mawr, Bedwas*; near Tre-hir Quarry, Bedwas*. **5.** Newport Docks*.

Hieracium anglorum (A. Ley) Pugsl.

Roadside banks; frequent. Native.
1. Govilon*, *Ley*. **2.** Machen, *Riddelsdell*. **4.** Wynd Cliff, *Pugsley*.

Hieracium strumosum (W. R. Linton) A. Ley

Banks; rare. Native.
3. Near Usk*, *Rickards*.

Hieracium acuminatum Jord.

Banks; rare. Native.
2. Pontypool, *Bladon*, (BM). **4.** Tintern, *Hort*, (BM).

Hieracium lachenalii C. C. Gmel.
 H. sciaphilum (Uechtr.) F. J. Hanb.

Banks, walls, railway banks and wood borders; common. Districts 2-5. Native.

Hieracium scabrisetum (Zahn) Roffey

Banks; rare. Native.
2. Pen-y-rhiw, Risca*.

Hieracium trichocaulon (Dahlst.) Roffey

Wood borders; rare. Native.
4. Wyes Wood, near Trelleck, *Shoolbred*.

Hieracium eboracense Pugsl.
 H. tridentatum sensu Pugsley *pro parte*

Banks; frequent. Native.
1. Near Pont Esgob; near Queen's Head, Honddu Valley, *Ley*. **3.** Near Brook Farm, Llanddewi Fach*. **4.** Chepstow, *Morgan:* Trelleck, *Ley*.

Hieracium calcaricola (F. J. Hanb.) Roffey
 H. tridentatum sensu Pugsley *pro parte*

Banks; frequent. Native.
1. Grwyne Fawr Valley, *Ley*. **2.** Aberbeeg. **3.** Near Cilfeigan*. **4.** Near Gaer Hill, near St. Arvans; near Trelleck, *Ley:* Wyes Wood, near Trelleck, *Shoolbred*.

COMPOSITAE

Hieracium umbellatum L.

Pastures, woods and banks; frequent to locally common. Districts 2-5. Native.

Hieracium perpropinquum (Zahn) Druce
H. boreale auct.

Woods, roadside banks and walls; common. Districts 1-4. Native.

Hieracium pilosella L.   Mouse-ear Hawkweed

Rough pastures, quarries, banks, open woods, walls, heaths and colliery tips; common. All districts. Native.

Hieracium brunneocroceum Pugsl.   Orange Hawkweed

Roadsides, walls and railway banks; rather frequent. Naturalized garden escape.
1. Kymin Hill, Monmouth*; near The Hendre*, *Charles:* Brook Estate, Rockfield. **2.** Ynysddu. **3.** Rogerstone, *Beal.* **4.** Chepstow, *Hort:* Tintern *Ley, Linton, Sandwith:* Shirenewton; between Itton School and Devauden; near Nine Wells, Tintern, *H. A. Evans.*

CREPIS L.

Crepis vesicaria L. ssp. **taraxacifolia** (Thuill.) Thell.
C. taraxacifolia Thuill.   Beaked Hawksbeard

Waste ground, railway banks, roadsides and hedgebanks; frequent to common. All districts. Colonist.

Crepis biennis L.   Rough Hawksbeard

Roadsides; rare. Native.
3. Near Raglan Castle, *Sandwith.* **4.** Near Crick*; Caldicot, *Shoolbred.*

Crepis capillaris (L.) Wallr.   Smooth Hawksbeard

Roadsides, cultivated and waste ground, meadows, pastures, walls, railway banks and heaths; common. All districts. Native.

TARAXACUM Weber

Taraxacum officinale Weber   Dandelion

Grassy places, waste and cultivated ground; very common. All districts. Native.

Taraxacum palustre (Lyons) DC.   Marsh Dandelion

Marshes, wet pastures and boggy places; frequent. Native.
1. The Ffwddog, Honddu Valley, *Ley:* Craig Syfyrddin, Cross Ash*,

COMPOSITAE

Charles. **2.** Near Blaen-y-cwm, Dukestown*; Trefil. **4.** Near Tintern; Shirenewton; Portskewett; Trelleck Bog; Kilgwrrwg; Devauden, *Shoolbred:* Chepstow, *Druce.* **5.** Marshfield, *Richards.*

Taraxacum laevigatum DC.
T. erythrospermum Andrz. ex Bess.
Roadsides, sandy and gravelly places and walls; frequent to locally common. All districts. Native.

MONOCOTYLEDONES

ALISMATACEAE

BALDELLIA Parl.

Baldellia ranunculoides (L.) Parl.   Lesser Water-plantain
Reens, swampy marshes and bogs; rare. Native.
1. Llanthony district, *Hamilton.* **3.** Near Llandenny, *Charles.* **4.** Trelleck Bog, *Ley.* **5.** Magor*, *Hamilton, Shoolbred:* near Rumney*.

ALISMA L.

Alisma plantago-aquatica L.   Water-plantain
Swampy marshes, reens, by streams and canals; common in the south and east of the county, locally so elsewhere. All districts. Native.

Alisma lanceolatum With.   Narrow-leaved Water-plantain
Marshes and reens; rare. Native.
5. Peterstone Wentlloog*, *Harrison:* Undy*.

SAGITTARIA L.

Sagittaria sagittifolia L.   Arrowhead
Reens and canals; locally common in District 5, rare elsewhere. Native.
2. Abercynon, *McKenzie.* **5.** St. Brides Wentlloog; Peterstone Wentlloog, *Hamilton,!:* Marshfield*, *Clark,!:* near Newport Lighthouse, *McKenzie:* Llanwern, *T. G. Evans.*

BUTOMACEAE

BUTOMUS L.

Butomus umbellatus L.   Flowering Rush
Reens and rivers; locally frequent. Native.
1. River Wye from Monmouth to Dixton*, *Charles.* **4.** St. Pierre, *Clark:* Penallt; Whitebrook*, *Charles.* **5.** Near Severn Tunnel Junction, *Lowe:* near Newport Lighthouse, *McKenzie:* Llandevenny, *Mrs. Ellis:* Peterstone Wentlloog*, *Harrison,!:* Magor, *T. G. Evans:* near Llanwern; by Pil-du Farm, St. Mellons*.

HYDROCHARITACEAE

HYDROCHARIS L.

Hydrocharis morsus-ranae L.   Frog-bit

Reens, ditches and ponds; rare to locally common. Native.
5. Marshfield*, *Clark,!:* Magor Moors*, *Hamilton, H. A. Evans, Shoolbred,!, T. G. Evans:* Llandevenny, *Mrs. Ellis:* Llanwern; Peterstone Wentlloog*; Rumney; near Whitson.

ELODEA Michx.

Elodea canadensis Michx.   Canadian Waterweed

Canals, reens, rivers and ponds; locally common. Denizen.
1. Abergavenny. **3.** Common in the Newport to Brecon Canal. **4.** River Wye, near Llandogo*; St. Pierre; Mounton; Pwllmeyric, *Shoolbred:* Fairoak Pond, *T. G. Evans,!.* **5.** Common in the reens.

First record: Shoolbred, 1893.

LAGAROSIPHON Harv.

Lagarosiphon major (Ridl.) Moss

Denizen. A native of South Africa.
2. Abundant in a quarry pool, near Gelligroes*, 1968.

JUNCAGINACEAE

TRIGLOCHIN L.

Triglochin palustris L.   Marsh Arrow-grass

Marshes; locally common. Native.
1. Abergavenny district, *White:* Honddu and Grwyne Fawr Valleys, *Ley:* Osbaston*, *Charles:* near Pont-y-spig*; Llanfoist*. **3.** Usk, *Clark:* near Ynys-y-fro; near Malpas Canal, *Hamilton:* Michaelstone y Vedw*. **4.** Trelleck Bog, *Watkins.* **5.** St. Brides Wentlloog; Peterstone Wentlloog, *Hamilton.*

Triglochin maritimum L.   Sea Arrow-grass

Salt marshes; common along the shore of the Severn and the tidal banks of the Rivers Usk and Wye. Districts 4 and 5. Native.

POTAMOGETONACEAE

POTAMOGETON L.

Potamogeton natans L.   Floating Pondweed

Canals, rivers, ponds and reens; locally frequent. Native.
1. Clytha, *Reader.* **3.** Usk, *Clark:* by Coed Mawr, near Rhiwderyn; canal,

POTAMOGETONACEAE

Newport to Bettws Road*. **4.** Mathern; St. Pierre; near Tintern, *Shoolbred:* Pen-y-parc, *T. G. Evans:* Penallt Common*. **5.** Peterstone Wentlloog*, *Harrison:* between Marshfield and Coedkernew*; Rumney*.

Potamogeton polygonifolius Pourr.   Oblong-leaved Pondweed
Peaty pools and streams; common in the upland districts, locally common elsewhere. Districts 1-4. Native.

Potamogeton lucens L.   Shining Pondweed
Rivers; very rare. Native.
1. River Wye, Monmouth*, *Charles.*

(**Potamogeton alpinus** Balb. Recorded in Watson, *New Botanists' Guide* on the authority of Conway. Probably an error, a form of *P. polygonifolius* having been mistaken for it.)

Potamogeton perfoliatus L.   Perfoliate Pondweed
Rivers, canals and reens; locally frequent. Native.
3. Canal near Llanfihangel Llantarnam*; canal, Goytre*; canal, Malpas.
4. River Wye, Brockweir and Llandogo, *Shoolbred:* Redbrook*, *Charles.*
5. Whitson*; Llanwern*; Goldcliff Pill.

Potamogeton pusillus L.
P. panormitanus Biv.
Canals and reens; locally frequent. Native.
2. Canal near Pontypool*, *Conway.* **4.** Wentwood Reservoir*, *Mrs. Parris.*
5. Between Whitson and Goldcliff*; between Llanwern and Whitson*; Marshfield*.

Potamogeton berchtoldii Fieb.   Small Pondweed
P. pusillus auct.
Canals and reens; rare. Native.
1. Near Llanfihangel, *Ley:* Redding's Enclosure*, *Charles:* near Troy Station, Monmouth*. **2.** Canal near Pontypool, *Conway:* canal, Crumlin, *Bentham.*

Potamogeton trichoides Cham. & Schlecht.
Reens; very rare. Native.
5. About 1 mile north of Whitson Church*, *Mrs. Welch.*

Potamogeton crispus L.   Curled Pondweed
Canals, reens and ponds; frequent. Native.
1. Llanfoist; Llantilio Crossenny. **2.** Canal above Pontypool*, *Conway:*
3. Usk, *Clark:* Castleton. **4.** Mathern*; St. Arvans Grange*; Penterry*;

POTAMOGETONACEAE

Broadwell, near Crick, *Shoolbred:* Pen-y-Parc; St. Pierre, *T. G. Evans.*
5. Ditches and reens by the Severn, *Shoolbred:* Peterstone Wentlloog*,
Harrison: Rumney; near Llanwern.

Potamogeton pectinatus L.   Fennel-leaved Pondweed
Reens and rivers; locally frequent. Native.
4. Mathern, *Shoolbred:* River Wye, Redbrook*, *Charles.* **5.** Between Rogiet
and Magor*, *Shoolbred:* Rumney to Marshfield*.

GROENLANDIA Gay
Groenlandia densa (L.) Fourr.   Opposite-leaved Pondweed
Potamogeton densus L.
Rivers; very rare. Native.
3. Usk, *Clark.* No speciment exists in Clark's Herbarium, the record must
therefore be considered a doubtful one.

RUPPIACEAE
RUPPIA L.
Ruppia maritima L.   Tassel Pondweed
R. rostellata Koch
Reens; rare. Native.
5. Below Undy, and between Undy and Magor*, *Shoolbred.*

ZANNICHELLIACEAE
ZANNICHELLIA L.
Zannichellia palustris L.   Horned Pondweed
Ponds and reens; locally frequent. Native.
3. Llanhennock*, *Mrs. Parris.* **4.** Near Crossway Green*, *Shoolbred:*
5. St. Pierre; Undy, *Shoolbred.*

var. pedicellata (Fr.) Wahlenb. ex Rosen.
Z. maritima Nolte ex G. F. W. Mey.
4. Brackish pond, Thornwell, Chepstow; reens below Mathern*,
Shoolbred. **5.** Rumney*, *Richards,!:* Magor, *T. G. Evans:* reens near
Undy*.

LILIACEAE
NARTHECIUM Huds.
Narthecium ossifragum (L.) Huds.   Bog Asphodel
Bogs, wet heaths and moors; locally common. Native.
2. Blorenge, *Hamilton:* Trefil*; Cwm Carno, near Brynmawr. **4.** Pen-y-
cae-mawr, near Wentwood, *Clark:* Trelleck Bog, *Ley, Hamilton, Shoolbred,
Charles,!:* above Tintern, *Shoolbred:* Pen-y-fan, near Whitebrook*,
Charles: near Tymawr, Penallt, *Beckerlegge:* Whitelye Common.

LILIACEAE

CONVALLARIA L.

Convallaria majalis L.   Lily of the Valley

Woods on calcareous soils; locally common. Native.

1. Near Skenfrith*, *T. H. Thomas, Charles:* Lilyrock Wood; near Hadnock Quarry*; railway bank between Dixton and Hadnock*; Rockfield; Coleford Road, Monmouth, *Charles.* **4.** Black Cliff Wood, *Gissing,!:* Wynd Cliff Wood*, *Clark, Ley, Shoolbred:* Tintern Woods*, *Hamilton, Richards:* near Chepstow; St. Arvans, *Hamilton:* The Barnetts; near Pandy Mill, Itton; Dinham; near Llanvair Discoed; Coppice Mawr, Itton*, *Shoolbred:* Brisca-bach Wood, Itton, *Lowe:* near St. Brides Netherwent*, *Mrs. Ellis.*

POLYGONATUM Mill.

Polygonatum odoratum (Mill.) Druce   Angular Solomon's Seal

Rocky woods on calcareous soils; very rare. Native.

4. Near Tintern*, *Purchas, Gissing, Ley, Shoolbred.*

Polygonatum multiflorum (L.) All.   Solomon's Seal

Woods; rare. Native.

1. Llanfihangel Crucorney, *Richards.* **3.** Between St. Mellons and Michaelstone y Vedw*, *Richards, Smith,!:* Plas Machen, *Campbell.* **4.** Llangwm, *Clark:* Wentwood*, *Ley, Hamilton, Shoolbred:* near Newbridge on Usk, *Lucus:* St. Pierre, *T. G. Evans.*

Hamilton's record for the Wynd Cliff is probably an error for *P. odoratum.*

ASPARAGUS L.

Asparagus officinalis L. ssp. **officinalis**   Asparagus

Waste places and railway sidings; rare. Alien.

1. Near Wyesham Signal Box, for several years, *Charles.* **5.** Newport, *Clark.*

RUSCUS L.

Ruscus aculeatus L.   Butcher's Broom

Woods and copses; very rare. ? Native.

3. Between Castleton and Michaelstone y Vedw, *Smith.* **4.** Between Bicca Common and The Coombe, *Dr. Clarke.*

LILIUM L.

Lilium martagon L.   Martagon Lily

Woods; very rare. Denizen.

4. Near Tintern*, *Shoolbred, Miss E. David:* near Chepstow*, *Miss Vachell.*

LILIACEAE

ORNITHOGALUM L.

Ornithogalum umbellatum L.   Star-of-Bethlehem

Pastures; very rare. Alien.

1. Near Monmouth*, *Richards.*

ENDYMION Dumort.

Endymion non-scriptus (L.) Garcke   Bluebell

Scilla non-scripta (L.) Hoffm. & Link

Woods and pastures by woods; common. All districts. Native.

COLCHICUM L.

Colchicum autumnale L.   Meadow Saffron

Pastures and open woods; locally frequent to locally common. Native.
1. Fairly common. **3.** Trevethin*, *Conway:* Trostrey, *Clark:* Raglan district, *Miss Frederick.* **4.** Chepstow*, *Conway, Clark, Hamilton, Shool-bred:* Rogiet*, *Hamilton, Shoolbred:* Mounton*; Itton*; Shirenewton; Dinham, *Shoolbred:* Caerwent, *Miss Vachell:* Pen-y-clawdd*, *Charles:* Ifton Wood, *Mrs. Ellis:* Cockshoot Wood, *T. G. Evans.*

Much less common than formerly owing to its having been eradicated from pastures as a noxious weed.

PARIS L.

Paris quadrifolia L.   Herb Paris

Woods; common in District 4, frequent in District 1, rare elsewhere. Native.
1. Abergavenny district, *White:* near Rockfield*, *Grimes:* Graig Syfyrddin, near Cross Ash*; Redding's Enclosure; Lady Park Wood; Hadnock Wood*; Osbaston Wood*; Staunton Road, Monmouth*, *Charles:* Seargent's Grove, near King's Wood*. **2.** Twyn-gwyn Dingle, Pontypool*; above Mamhilad, *T. H. Thomas.* **4.** Common.

First recorded by John Ray in 1662 from near Tintern.

JUNCACEAE

JUNCUS L.

Juncus squarrosus L.   Heath Rush

Common and often abundant on moors, wet heaths and bogs. Districts 1, 2 and 4. Native.

Juncus tenuis Willd.   Slender Rush

Very rare. Denizen.
2. Damp ground by an old colliery tip, Crosskeys*, 1968.

Juncus compressus Jacq.　Round-fruited Rush

Marshy places; very rare. Native.

1. Near White House Farm, Llanvihangel Ystern Llewern, *Milne-Redhead*.

Juncus gerardii Lois.　Gerard's Rush

Salt marshes and tidal banks of rivers; common. Districts 3-5. Native.

Juncus bufonius L.　Toad Rush

Damp roadsides and cart-tracks, and marshy places; common. All districts. Native.

Juncus inflexus L.　Hard Rush

Wet pastures and roadsides on heavy soils; common, especially on the marine alluvium. All districts. Native.

Juncus effusus L.　Soft Rush

Wet pastures, marshes, wet heaths, bogs and moors, favouring wetter and more acid habitats than *J. inflexus:* common. All districts. Native.
The var. **compactus** Hoppe is abundant on the moorlands.

×**inflexus** (*J.* × *diffusus* Hoppe)
5. Near New House, Rumney.*

Juncus subuliflorus Drej.　Dense-headed Rush
　Juncus conglomeratus auct. non L.

Wet pastures and heaths, marshes and bogs; locally common to frequent. Districts 1, 3-5. Native.

Juncus maritimus Lam.　Sea Rush

Salt marshes; rare. Native.
5. Near St. Pierre Pill*, *Shoolbred.*

Juncus subnodulosus Schrank　Blunt-flowered Jointed Rush
　J. obtusiflorus Ehrh. ex Hoffm.

Acid marshes; very rare. Native.
4. Between Trelleck and Penallt, *Ley.*

Juncus acutiflorus Ehrh. ex Hoffm.　Sharp-flowered Jointed Rush
　J. sylvaticus auct.

Marshes and wet woods; common. All districts. Native.

JUNCACEAE

Juncus articulatus L. Jointed Rush
Marshes, wet heaths, moors and pond margins; common. All districts.
Native.

Juncus bulbosus L. Bulbous Rush
J. supinus Moench
Moors, wet heaths and bogs; common in District 2 and the upland parts
of District 1, less common elsewhere and absent from District 5. Native.

LUZULA DC.

Luzula pilosa (L.) Willd. Lesser Woodrush
Woods and shady hedgebanks; common. Districts 1-4. Native.

Luzula forsteri (Sm.) DC. Forster's Woodrush
Woods, mainly on calcareous soils; locally common to locally frequent.
Native.
1. Common in the woods about Monmouth: near the Queen's Head,
Honddu Valley, *Ley:* Fforest, near Abergavenny*, *Charles:* **4.** Piercefield
Woods, *Hamilton, Shoolbred:* Mounton; Barbadoes Hill, Tintern*; Fair
Oak, *Shoolbred:* Whitebrook*; Cuckoo Wood, Llandogo*; Trelleck
Grange, near Tintern Cross*, *Charles:* Wynd Cliff and Black Cliff Woods.

×**pilosa** (*L.* ×*borreri* Bromf. ex Bab.)
4. Near Lady Mill, Mounton*; Barbadoes Hill, Tintern; between Wynd
Cliff and Tintern, *Shoolbred.*

Luzula sylvatica (Huds.) Gaudin Great Woodrush
Woods and shady river banks; frequent to locally common. Districts 1-4.
Native.

var. **gracilis** Rostrup
4. Tintern, *Amherst.*

Luzula luzuloides (Lam.) Dandy & Wilmott
L. nemorosa (Poll.) E. Mey.
Shady banks; very rare. Denizen.
4. Between Wynd Cliff and Tintern, *Miss M. Cobbe.*

Luzula campestris (L.) DC. Field Woodrush
Pastures, grassy banks and roadsides; common. All districts. Native.

Luzula multiflora (Retz.) Lejeune Many-flowered Woodrush
L. erecta Desv.
Heaths and woods; frequent. Districts 1-4. Native.

AMARYLLIDACEAE

ALLIUM L.

Allium vineale L.   Crow Garlic

Pastures, hedgebanks and railway banks; rare to locally frequent. Native.
1. Between Monmouth and Hadnock; between Monmouth and Dixton, *Charles:* near Wyesham*, *R. Lewis, Charles.* **3.** Llandenny*, *Miss Davis:* Caerleon. **4.** Rogiet, *Archdeacon Bruce:* Portskewett, *Shoolbred, Marshall:* Chepstow; Tintern*, *Shoolbred, Miss Francis:* Mathern*, *Shoolbred.* **5.** Magor Pill, *Shoolbred:* Undy; Mathern Pill, *T. G. Evans.*

Allium oleraceum L.   Field Garlic

Hedgebanks and roadsides; rare. Denizen.
4. Portskewett*, *Shoolbred, R. Lewis:* Dewstow Lane, near Caldicot*, *Shoolbred:* near Mounton, *Mrs. Macnabb:* near Crick, *R. Lewis.*

Clark in his *Flora of Monmouthshire* gives records from Pontsampit, near Usk and Caerleon, but these probably refer to *Allium vineale* which he omits.

Allium schoenoprasum L.   Chives

Garden escape or relic of cultivation.
1. Field on the Kymin, Monmouth*, 1943, *Charles.*

Allium roseum L. ssp. **bulbiferum** (DC.) E. F. Warburg

Alien.
1. Llanover*, 1951, *Freer.*

Allium ursinum L.   Ramsons

Woods; common on the limestone and locally common on the Old Red Sandstone. Districts 1-4. Native.

LEUCOJUM L.

Leucojum acstivum L.   Summer Snowflake

Alien.
3. St. Mellons Golf Course*, 1966, *R. Williams.*

GALANTHUS L.

Galanthus nivalis L.   Snowdrop

Wet woods and copses, hedgebanks and river banks; locally frequent. Denizen.
1. Beaulieu Wood*; Hadnock*; Skenfrith; Norton; Cross Ash*; Newcastle; Rockfield; Garth Wood; Kymin Hill*; banks of the Wye, Monmouth, *Charles.* **2.** Near The Folly, Pontypool; above Nant-y-gollen, *T. H. Thomas.* **3.** Between St. Mellons and Michaelstone y Vedw*,

199

AMARYLLIDACEAE

Miss Vachell, Smith,!: Ffrwd Brook, between Llanhennock and Llandegfedd parishes; near Walnut Tree, Llandegfedd; near Plas Machen, *Campbell:* Glascoed Common*, *Briggs:* Michaelstone Bridge*. **4.** Near Shirenewton*; Itton, *Hamilton, Shoolbred:* near Tintern; Earlswood*, *Shoolbred:* above Broadwells Farm, near Pwllmeyric, *Willan:* Pen-y-fan, Whitebrook*; Cuckoo Wood, Llandogo*; Lone Lane, Pentwyn; Newmills, Penallt*, *Charles:* St. Brides Netherwent, *Mrs. Ellis.* **5.** Marshfield; Pike.

NARCISSUS L.

Narcissus pseudonarcissus L.   Wild Daffodil

Woods and pastures; widespread and often locally common in Districts 1, 3 and 4, unrecorded from Districts 2 and 5. Native.

Narcissus hispanicus Gouan
   N. major Curt.

Rare. Denizen.
3. Stream bank, Llanrumney, *Norton.*

Narcissus majalis Curt.   Pheasant's-eye Narcissus
   N. poeticus auct.

Pastures and woods; rare. Denizen.
1. Govilon, a double-flowered form, *T. H. Thomas.* **3.** In several places about Usk, *Clark.* **4.** Near Chepstow, *Bossey:* near Shirenewton; St. Arvans, *Shoolbred:* Runston*, *Lowe, Shoolbred.*

Narcissus × **biflorus** Curt.   Two-flowered Narcissus

Pastures, old orchards and woods; rare. Denizen.
1. Newton*, *Charles.* **3.** Near Pentwyn and Gwehelog, *Clark:* near the Maypole Inn, between Church Road and Rhiwderyn, *Rönnfeldt.* **4.** Near Pen-y-clawdd, *Rep. Woolhope Club,* 1885; old orchards and pastures in several places, *Shoolbred.*

IRIDACEAE
IRIS L.

Iris foetidissima L.   Gladdon

Woods on the carboniferous limestone; rare. Native.
4. Near Portskewett; Dinham; near Shirenewton, *Shoolbred:* Thicket Wood, near Rogiet, *Clarke:*

Iris pseudacorus L.   Yellow Flag

Marshes, pond and canal margins; common in all districts except 2. where it is rare. Native.

IRIDACEAE

CROCUS L.

Crocus sativus L.   Saffron

Alien.

4. Field near Rogiet Rectory 1903, *Bennett*.

CROCOSMIA Planch.

Crocosmia × **crocosmiflora** (Lemoine) N.E.Br.   Montbretia

Established garden escape.

2. Roadside, near Gelligroes, 1968. **4.** Black Cliff, *T. G. Evans*.

DIOSCOREACEAE

TAMUS L.

Tamus communis L.   Black Bryony

Hedges and wood borders; common. All districts. Native.

ORCHIDACEAE

CEPHALANTHERA Rich.

Cephalanthera longifolia (L.) Fritsch   Narrow-leaved Helleborine

Woods on the limestone; very rare. Native.
4. Between Chepstow and Tintern*, *Lightfoot, Shoolbred, Miss Vachell,
T. G. Evans*.

First recorded by John Lightfoot in 1773.

EPIPACTIS SW.

Epipactis palustris (L.) Crantz   Marsh Helleborine

Marshes; very rare. Native.
3. Near Castell-y-bwch, Henllys*, *H. Rowland.!.*

Epipactis helleborine (L.) Crantz   Broad-leaved Helleborine
   E. latifolia (L.) All.

Woods; locally frequent. Native.
1. Abergavenny district, *White:* near Rockfield, *Grimes:* Lilyrock Wood;
Garth Wood*; between The Kymin and Beaulieu Farm, Monmouth;
Hadnock Wood*, *Charles*. **2.** Cwm Ffrwd-Oer, near Abersychan, *Vernall:*
Coed-y-person, near Llanfoist, *Guile*. **3.** Llangybi Woods†, *Clark:* Llan-
tarnam Abbey, *Horne*. **4.** Near Tintern*, *Purchas:* Whitebrook, *Ley:*
Forester's Oak, near St. Arvans, *Marshall & Shoolbred:* Castle Woods,
Chepstow*; Piercefield; St. Arvans Grange; The Narth*; Portskewett;
Wentwood, *Shoolbred:* Llanmelin*, *Miss Cooke:* Little Llanthomas,
Miss M. Parry: Pen-y-clawdd*, *Miss Beavan:* Kilpale, Shirenewton,
Mrs. Ellis: Wynd Cliff*; Black Cliff, *Shoolbred, T. G. Evans.!.*

ORCHIDACEAE

Epipactis leptochila (Godfery) Godfery   Green-flowered Helleborine
Woods, very rare. Native.
4. Between the Wynd Cliff and Tintern, *C. E. Salmon.*

SPIRANTHES Rich.

Spiranthes spiralis (L.) Chevall.   Autumn Lady's-tresses
Dry limestone banks, pastures and quarries; locally frequent. Native.
1. Abergavenny district, *White:* Honddu Valley, *Ley:* near Greenmoors, Llanddewi Ysgyryd*, *Willan:* Beaulieu Farm, Monmouth; near Manson's Cross*; Hadnock Farm, near Monmouth*; Ysgyryd Fawr*; Upper Redbrook*; near Halfway House Wood, Monmouth*, *Charles:* Buckholt Wood, *Lane.* **2.** Pen-y-garn, Pontypool*, *T. H. Thomas.* **3.** Bettws Newydd†; Llangybi Castle†, *Clark:* Usk, *Rickards.* **4.** The Barnetts Farm*; Dinham; Shirenewton*; Earlswood; Wye's Wood Common; Rogiet; Little Dinham; near St. Brides Netherwent*; Trelleck*, *Shoolbred:* Bayfield, near Chepstow, *Hamilton:* Rogiet, *Bennett, Mrs. Ellis:* Pen-y-clawdd, *Miss Davies:* Mounton, *T. G. Evans.* **5.** Marshfield*, *Conway.*

LISTERA R. Br.

Listera ovata (L.) R. Br.   Twayblade
Woods and boggy places; common in the Wye Valley, frequent elsewhere. Districts 1-4. Native.

NEOTTIA Ludw.

Neottia nidus-avis (L.) Rich.   Bird's-nest Orchid
Woods, usually about the roots of hazel; frequent in District 4, rare elsewhere. Native.
1. Wye Valley above Monmouth, *Hamilton:* Lilyrock Wood; Beaulieu Wood*; Wyesham*; Lady Park Wood*; Hadnock Wood, *Charles.* **2.** Lasgarn Wood, Abersychan, *Miss Jones.* **3.** Raglan Castle, *Gissing:* Bath Wood, Llanfihangel Llantarnam, *Horne.* **4.** Between Monmouth and Tintern, *Gissing:* near Tintern, *Hamilton:* Castle Wood, Chepstow; St. Arvans; Dinham; The Grondra, *Shoolbred:* Tintern Woods*, *Richards:* Graig Wood, Pen-y-clawdd*, *Miss Davis:* near Treveldy Farm, Pen-y-clawdd, *Miss Beavan:* Thicket Wood, near Rogiet, *Mrs. Ellis,!:* Minnetts Wood, *Horne:* Slade Wood, *H. Rowland:* The Barnetts*; Wynd Cliff Woods, *Shoolbred, T. G. Evans:* St. Pierre's Great Woods, *T. G. Evans:* Black Cliff Wood*.

COELOGLOSSUM Hartm.

Coeloglossum viride (L.) Hartm.   Green Frog Orchid
Habenaria viridis (L.) R. Br.
Meadows and pastures; rare. Native.

ORCHIDACEAE

1. Near Cwmyoy, *Dr. Wood:* Grwyne Fawr Valley, *Ley:* near the 'Viaduct', Monmouth, *Charles.* **4.** Tintern, *anon, Reader:* near Tintern, *Gissing:* Usk Road, near Chepstow, *Hamilton:* Barnett Woods*; near Chepstow*; Shirenewton*; Dinham*; The Mount, Chepstow*, *Shoolbred:* Hangman's Wood, Shirenewton, *Lowe:* Cwmcarvan*, *Miss M. Parry:* Pen-y-clawdd*, *Miss Davis.* **5.** St. Brides Wentlloog Road, between Ebbw Bridge and Pheasant Bridge, *Hamilton.*

GYMNADENIA R. Br.

Gymnadenia conopsea (L.) R. Br.   Fragrant Orchid
Habenaria conopsea (L.) Benth. non Reichb. f.

Bogs and peaty marshes; rather rare. Native.
1. Abergavenny district, *White:* Grwyne Fawr Valley, *Ley:* Honddu Valley, *Thompson, Sandwith:* north of Abergavenny, *Hamilton.* **3.** Hill Farm, Trostrey, *Clark:* Trostra Farm, south of Pontypool*, *T. H. Thomas:* near Coed-y-paen, *Rickards:* near Castell-y-bwch, Henllys*. **4.** Newchurch*; near the Goitre, Chepstow, *Shoolbred:* near Chepstow, *Hamilton, Miss Frederick:* Little Llanthomas*, *Miss M. Parry:* Trelleck Bog.

PSEUDORCHIS Séguier

Pseudorchis albida (L.) A. & D. Löve.   White Frog Orchid
Habenaria albida (L.) R. Br.

Meadows; very rare. Native.
1. Head of the Grwyne Fawr Valley, *Ley:* Coed-dias, Grwyne Fawr Valley*, *C. T. & E. Vachell.*

PLATANTHERA Rich.

Platanthera chlorantha (Custer) Reichb.   Greater Butterfly Orchid
Habenaria chlorantha (Custer) Bab. non Spreng.

Woods, shady banks and pastures; locally frequent. Native.
1. Abergavenny district, *White:* near Rockfield, *Grimes:* Pentre, Llangattock Vibon Avel*; Lady Park Wood; between Monmouth and the Bibblings; near The Slaughter, within the county boundary; Halfway House Wood; between Llanthony and Capel-y-ffin*; near Monmouth*, *Charles:* near Tregare*. **3.** Near Coed-y-paen, *Rickards.* **4.** near Tintern, *Gissing:* Shirenewton; Wynd Cliff*; Kilgwrrwg; Rogiet; Mounton; between Trelenny and the Grondra*, *Shoolbred:* Chepstow Park*, *Rickards:* between Monmouth and Redbrook, *Charles:* Pen-y-clawdd*; *Miss Davis:* Thicket Wood, Ifton Wood, and The Minnetts, Rogiet, *Mrs. Ellis,!:* St. Pierre's Great Woods; Llangwm, *T. G. Evans.*

Platanthera bifolia (L.) Rich.   Lesser Butterfly Orchid
Habenaria bifolia (L.) R. Br.

Heaths, pastures and woods; locally frequent. Native.

ORCHIDACEAE

1. Grwyne Fawr Valley, *Ley:* Abergavenny district, *White:* Coed-dias, Grwyne Fawr Valley, *C. T. & E. Vachell:* above Llanthony*, *Charles:* near Grosmont, *Sankey-Barker.* **2.** The Dingle, Pontypool*, *T. H. Thomas.* **4.** Rogiet, *Hamilton, Bennett, Shoolbred:* near Llanmelin; Barnett Woods*; Mounton*; near St. Arvans; Shirenewton, *Shoolbred:* Itton, *Lowe:* Trelleck Bog*, *Rickards:* Tymawr, Lydart*; Penallt Common, *Beckerlegge:* Slade Wood, *H. Rowland:* Pen-y-fan, near Whitebrook*, *Charles.* Hamilton's unconfirmed record from Penmoel is omitted since he did not distinguish between *P. chlorantha* and *P. bifolia.*

OPHRYS L.

Ophrys apifera Huds.   Bee Orchid

Pastures, banks and quarries on calcareous soils; rare to locally frequent· Native.
1. Near Rockfield, *Grimes:* railway bank by Lady Park Wood*, *Charles:* Wyesham*, *Miss Sainsbury.* **2.** Risca. **3.** Llangibby*; near Coed-y-paen, *Rickards.* **4.** Chepstow*, *Lightfoot, E. Lees, Shoolbred:* between Tintern and Wynd Cliff, *Ley:* Rogiet, *Hamilton:* near Tintern; Fairfield Farm, Chepstow*, *Shoolbred:* Wentwood, *Miss Frederick:* The Minnetts and Slade Meadow, Rogiet, *Mrs. Ellis:* Ifton, *Horne:* Mounton, *T. G. Evans:* Langstone*. **5.** Near the mouth of the River Usk, Newport*, *Miss Collins.*

First recorded by John Lightfoot in 1773.

Ophrys insectifera L.   Fly Orchid
O. muscifera Huds.

Woods on calcareous soils; rare. Native.
1. Lilyrock Wood; Lady Park Wood, *Charles.* **4.** Chepstow*, *Lightfoot, Shoolbred:* Wynd Cliff, *Shoolbred, Hamilton:* The Park, Shirenewton Hall, *Lowe:* Common Coed, near Undy, *Horne.*

First recorded by John Lightfoot in 1773.

ORCHIS L.

Orchis morio L.   Green-winged Orchid

Pastures and meadows; frequent. All districts. Native.

Orchis mascula (L.) L.   Early Purple Orchid

Woods; frequent. Districts 1-4. Native.

DACTYLORHIZA Nevski.

Dactylorhiza fuchsii (Druce) Soó ssp. **fuchsii**   Spotted Orchid
Orchis fuchsii Druce

Damp pastures, marshes and woods; common on the more or less calcareous soils of Districts 1, 3-5. Native.

ORCHIDACEAE

Dactylorhiza maculata (L.) Soó ssp. **ericetorum** (Linton) P. F. Hunt & Summerh.   Early Spotted Orchid
Orchis ericetorum (E. F. Linton) E. S. Marshall
Wet heaths, bogs and peaty marshes; common on acid soils. Districts 1, 3 & 4. Native.

× **praetermissa**
1. Llanfoist*. **4.** Near the Virtuous Well, Trelleck*.

Dactylorhiza incarnata (L.) Soó ssp. **incarnata**   Crimson Marsh Orchid
Orchis strictifolia Opiz
Rare. Native.
1. Boggy field near Pont-y-spig*.

Dactylorhiza praetermissa (Druce) Soó   Marsh Orchid
Orchis praetermissa Druce; *O. latifolia* auct.
Marshes, wet pastures and bogs; rather rare. Native.
1. Osbaston*, *Charles:* Pont-y-spig*. **3.** Coed-y-paen*, *Conway, Rickards:* Bettws Newydd*, *Rickards.* **4.** Near Brockweir, *Hamilton:* near St. Arvans*, *Hamilton, Shoolbred:* near Tintern, *Shoolbred:* Chepstow Park, *Rickards:* near Pen-y-clawdd, *Miss Davis:* Trelleck Bog, *Mrs. Ellis:* near The Virtuous Well, Trelleck*.

ANACAMPTIS Rich.
Anacamptis pyramidalis (L.) Rich.   Pyramidal Orchid
Orchis pyramidalis L.
Pastures and quarries on calcareous soils; rare. Native.
1. Near Hendre, Monmouth*, *Vernall.* **4.** Ifton Quarry, *Clarke:* between Chepstow and Tintern, *Hamilton:* near Portskewett*, *Marshall & Shoolbred:* by Caerwents Brook, between Caerwent and Dewstow*, *Mrs. Ellis:* Wynd Cliff; Black Cliff Wood; Caerwent Quarry, *T. G. Evans.*

ARACEAE
ARUM L.
Arum maculatum L.   Cuckoo Pint
Hedgebanks and woods; common. All Districts. Native.

LEMNACEAE
LEMNA L.
Lemna polyrhiza L.   Many-rooted Duckweed
Ponds, canals and reens; locally common. Native.
3. Canal from Newport to Abergavenny*. **4.** Mare's Pool, St. Arvans; St. Pierre Pond; between Crick and Mount Ballan, *Shoolbred.* **5.** Rumney to Marshfield*; Peterstone Wentlloog; Magor; Goldcliff Pill.

LEMNACEAE

Lemna trisulca L.   Ivy-leaved Duckweed
Reens, canals and ponds; common. Districts 1, 3-5. Native.

Lemna minor L.   Lesser Duckweed
Ponds, canals and reens; very common. All districts. Native.

Lemna gibba L.   Gibbous Duckweed
Ponds and reens; rare to locally frequent. Native.
4. Mare's Pool, St. Arvans, *Shoolbred:* Chepstow, *H. A. Evans:* Shire-newton, *Lowe.* **5.** Reens below Rogiet*, *Shoolbred:* Rumney*, *Miss Vachell:* Llanwern, *T. G. Evans:* Magor; Peterstone Wentlloog*.

SPARGANIACEAE
SPARGANIUM L.

Sparganium erectum L. ssp. **erectum**   Branched Bur-reed
  S. ramosum Huds.

Pond margins, streams, canals and reens; frequent to common. Districts 1, 3-5. Native.

ssp. **neglectum** (Beeby) Schinz & Thell.
1. Llantilio Crossenny. **2.** Canal, Risca to Crosskeys*. **3.** Near Pont Waun-y-pwll, Llanddewi Fach. **4.** Organy Pool*.

ssp. **microcarpum** (Neum.) Hylander
5. Reen near Pîl-du Farm, St. Mellons*.

Sparganium emersum Rehm.   Unbranched Bur-reed
  S. simplex Huds. *pro parte*

Reens, rivers, swampy places and canals; rare. Native.
1. River Wye, Monmouth*, *Charles.* **3.** Near Bassaleg*; near Rogerstone*. **4.** Earlswood, *T. G. Evans.* **5.** Peterstone Wentlloog*; Whitson.

TYPHACEAE
TYPHA L.

Typha latifolia L.   Reed Mace
Ponds, streams and reens; frequent to locally common. All districts. Native.

CYPERACEAE
ERIOPHORUM L.

Eriophorum angustifolium Honck.   Narrow-leaved Cotton-grass
Common in boggy places, wet moorland and acid marshes. Districts 1-4. Native.

Eriophorum latifolium Hoppe   Broad-leaved Cotton-grass
  E. paniculatum Druce
Boggy places; rare. Native.
1. Honddu and Grwyne Fawr Valleys; the Ffwddog, *Ley*. **2.** Varteg, *Clark:* Trefil*. **3.** Near Castell-y-bwch, Henllys*. **4.** Near Mitchel Troy, *Ley:* Trelleck Bog, *Watkins,!*.

Eriophorum vaginatum L.   Hare's-tail Cotton-grass
Common in boggy places, wet moorlands and acid marshes. Districts 1, 2 & 4. Native.

SCIRPUS L.

Scirpus caespitosus L.   Deerhair Sedge
  Trichophorum caespitosum (L.) Hartm.
Very common on wet moorland and boggy places. Districts 1-4. Native.

Scirpus maritimus L.   Sea Clubrush
Common in salt marshes and reens by the Severn, and on the tidal banks of the rivers. Districts 4 & 5. Native.

Scirpus sylvaticus L.   Wood Clubrush
Marshes and river banks; rare. Native.
1. Osbaston*; Upper Redbrook*; Newton Dixton*, *Charles*. **3.** Garcoed; Olway Brook, Usk, *Clark:* Michaelstone Bridge, Michaelstone y Vedw*; near Plas Machen*. **4.** Near Pont y Saeson, *Mrs. Welch:* Mitchel Troy*; River Wye, Penallt*, *Charles*.

Scirpus tabernaemontani C. C. Gmel.   Bulrush
Reens; rare. Native.
5. Near Rogiet, *Shoolbred*:* Marshfield, *Richards*.

Scirpus setaceus L.   Bristle Clubrush
  Isolepis setacea (L.) R. Br.
Marshes and pond margins; widespread but not common. Native.
1. Near Llanthony*, *Ley, Charles:* Hadnock Wood*, *Charles:* Llanfoist. **2.** Near the Punchbowl, Blorenge*; Cwm Gwyddon, Abercarn. **3.** Near Pont-y-clivon, near Usk†, *Clark:* Christchurch*. **4.** Below Mathern; Barbadoes Hill, Tintern, *Shoolbred:* Trelleck Bog, *Charles*. **5.** Near Caldicot, *Shoolbred*.

ELEOCHARIS R. Br.

Eleocharis palustris (L.) Roem. & Schult. agg.   Marsh Clubrush
Frequent in marshes and on the margins of ponds. All districts. Native.

CYPERACEAE

Eleocharis uniglumis (Link) Schult.

Rare. Native.

2. Wet peaty ground, Machen*.

(Clark's records for *Eleocharis multicaulis* (Sm.) Sm. and *Eleocharis acicularis* (L.) Roem. & Schult. are probable errors since unlocalised specimens in his Herbarium so named are *Eleocharis palustris* and *Scirpus caespitosus* respectively.)

RHYNCHOSPORA Vahl

Rhynchospora alba (L.) Vahl   White Beakrush

Very rare. Native.

4. Trelleck Bog*, *Ley, Watkins, Shoolbred, Redgrove,!.* Formerly plentiful but now almost extinct due to the drying up of the bog.

CAREX L.

Carex laevigata Sm.   Smooth-stalked Sedge

Woods and wet heaths; frequent in District 4, rare elsewhere. Native.
1. Abergavenny district, *White:* near Pont-y-spig*; Grwyne Fawr Valley, *Ley.* **3.** Near Pontnewydd*. **4.** Pen-y-cae-mawr, Wentwood†; Troggy Castle, *Clark:* Trelleck Bog, *Ley, Shoolbred:* Catbrook Lane and Yellow Moor, near Tintern*; Chepstow Park; Coetgae, Shirenewton*; New-church West; St. Arvans*; Kilgwrrwg Bottom; The Glynn, Itton, *Shoolbred:* near Bicca Common, near Castle Troggy Brook*, *J. N. Davies.*

Carex distans L.   Distant-spiked Sedge

Salt marshes and meadows subject to frequent inundation by the tide; locally frequent to common by the Severn and on tidal banks of the Usk and Wye. Districts 3-5. Native.

Carex hostiana DC.   Tawny Sedge

Boggy places and marshes; rare to locally frequent. Native.
1. Abergavenny district, *White:* Ffwddog; Olchon Dingle, Honddu Valley; near Llangattock Vibon Avel*, *Ley:* Black Mountains, above Llanthony*, *Charles:* near Pont-y-spig*; Llanfoist*. **2.** Trefil*. **3.** Near Castell-y-bwch*. **4.** Trelleck Bog, *Watkins:* between Trelleck and Penallt, *Ley:* Rhyd-y-fedw, Itton*; Kilgwrrwg Bottom*, *Shoolbred:* Pen-y-fan, Whitebrook*, *Charles.*

Carex binervis Sm.   Green-ribbed Sedge

Common on moorlands and in boggy places. Districts 1-4. Native.

Carex lepidocarpa Tausch
Marshy places on base rich soils; rare. Native.
1. Pont-y-spig*. **2.** Trefil*.

Carex demissa Hornem.   Yellow Sedge
Common in marshy and boggy places. Districts 1-4. Native.
(Clark's records for *C. flava* and *C. serotina* are errors.)

Carex extensa Gooden.   Long-bracted Sedge
Very rare. Native.
5. Salt marsh near Magor Pill*, *E. Nelmes,!.*

Carex sylvatica Huds.   Wood Sedge
Woods and shady hedgebanks; common. Districts 1-4. Native.

Carex pseudocyperus L.   Cyperus Sedge
Margins of ponds and reens; rare. Native.
1. Llantilio Crossenny*, *W. Nelmes.* **4.** The Barnetts Farm, Chepstow*;
St. Pierre Pond, *Shoolbred.* **5.** Reens between Mathern and St. Pierre,
Shoolbred: near Magor*, *E. Nelmes:* Near New House, Rumney*.

Carex rostrata Stokes   Beaked Sedge
Marshes and bogs; rare. Native.
1. Grwyne Fawr Valley; near Pont-y-spig, *Ley.* **2.** Cwm-carno-eilw, Brynmawr*. **4.** Trelleck Bog*, *Ley, Shoolbred, T. G. Evans:* Suck-pont Reservoir, *Shoolbred:* Fairoak Pond, *Shoolbred*,!:* Pen-y-fan, near Whitebrook*, *Charles:* near The Virtuous Well, Trelleck*.

Carex vesicaria L.   Bladder Sedge
Marshes, marshy thickets and stream-sides: rare. Native.
1. Grwyne Fawr Valley, *Ley:* Pont-y-spig*, *Ley, Wilmott,!:* Coalpit Hill,
Fforest, near Abergavenny*, *Charles:* Llanfoist*. **3.** Near Rhadyr Mill,
Usk†, *Clark:* near Plas Machen*. **4.** Trelleck Bog, *Watkins, Shoolbred:*
Mathern, *Shoolbred.*

Carex riparia Curt.   Great Pond Sedge
Reens, riversides and swampy places; locally common. Native.
1. Coalpit Hill, Fforest, near Abergavenny, *Charles.* **3.** Penpergwm,
Campbell. **4.** Below Tintern*; Fish Pond near The Grondra; St. Pierre
Pond; Shirenewton*; between Mathern and St. Pierre*, *Shoolbred.*
5. Common in reens by the Severn.

CYPERACEAE

Carex acutiformis Ehrh.   Lesser Pond Sedge

Reens, riversides and swampy places; locally common. Native.
1. Grwyne Fawr Valley, *Ley:* Fforest, near Abergavenny*, *Charles:* near Pont-y-spig*. **3.** Plas Machen. **4.** Rhyd-y-fedw, Shirenewton*; between Bigsweir and Brockweir, etc. *Shoolbred.* **5.** Common along reens by the Severn.

Carex pendula Huds.   Pendulous Sedge

Damp woods, lanes and streamsides; common in Districts 1 and 4, unrecorded elsewhere. Native.

Carex strigosa Huds.   Broad-leaved Sedge

Damp woods; locally frequent. Native.
1. Near Pont-y-spig, *Ley:* Crown Woods near Monmouth*, *Charles:* Lady Park Wood*; Redding's Enclosure*, *Charles, R. Lewis, Seddon,!:* Hadnock Wood*, *Charles.* **4.** Wynd Cliff, *Ley, Shoolbred:* Piercefield*, *Hamilton, Shoolbred:* Mounton, *Hamilton:* near Rogerstone Grange*; between Tintern and Penterry*; Shirenewton, *Shoolbred:* near Pen-y-clawdd, *Sandwith:* Upper Redbrook*, *Charles.*

Carex pallescens L.   Pale Sedge

Wet meadows, marshes and wet woods; locally frequent. Native.
1. Grwyne Fawr Valley*, *Miss Vachell:* Llangattock Vibon Avel*; Buckholt Wood*; Lady Park Wood*; Wonastow*; Hadnock Wood*, *Charles:* Llanfoist*; near Triley Bridge, near Abergavenny*. **3.** Cefnila Farm, near Usk, *Clark:* Raglan. **4.** Mounton*; Trelleck Bog; Rhyd-y-fedw; near Brooklands, Shirenewton; between Tintern and Trelleck*; Coetgae, Shirenewton, *Shoolbred:* near Penallt Common*, *Charles:* near Pen-y-clawdd, *Sandwith:* Cockshoot Wood, *T. G. Evans.*

Carex panicea L.   Carnation Sedge

Wet meadows, marshes and boggy places; common. Districts 1-4. Native.

Carex flacca Schreb.   Glaucous Sedge

Wet roadsides and banks, limestone quarries and calcareous grassland; common. All districts. Native.

Carex hirta L.   Hairy Sedge

Grassland, marshes, open woods and river banks; common. All districts. Native.

Carex pilulifera L.   Pill Sedge

Wet heaths, boggy places and woodland rides; locally frequent. Native.
1. Kymin, Monmouth*, *Ley, Charles:* Ruffet; Buckholt Wood*; near Fiddler's Elbow, near Monmouth*; Hadnock Wood*; Lady Park Wood*, *Charles:* near King's Wood, Monmouth*; Ysgyryd Fawr*. **2.** Below Twmbarlwm, Risca*. **3.** Pontnewydd, *Conway.* **4.** Between Itton Common and the Shepherd's Cottage*, *Purchas:* Trelleck Bog*; Yellow Moor, Tintern; Chepstow Park; Newchurch West, *Shoolbred:* Pen-y-fan, White-brook*; The Narth, near Trelleck*, *Charles:* Wentwood, *Sandwith.*

Carex caryophyllea Latourr.   Vernal Sedge

Heaths, pastures, grassy roadsides and railway banks; common. Districts 1-4. Native.

Carex montana L.   Hill Sedge

Rough grassy places on calcareous soils; locally frequent. Native.
4. Wynd Cliff*, *Purchas, Watkins, Ley, Shoolbred, T. G. Evans.*

Carex digitata L.   Fingered Sedge

Roadside banks, woodland rides and railway banks on calcareous soils; locally frequent. Native.
1. Lady Park Wood*; Redding's Enclosure*, *Charles.* **4.** Wynd Cliff*, *Purchas, Watkins, Shoolbred, Ley,!:* near Tintern Abbey, *Bailey:* Black Cliff Wood*.

Carex acuta L.   Slender-spiked Sedge

Riversides and marshy places; locally frequent to locally common. Native.
1. Banks of the River Wye between Dixton and Monmouth*; Hadnock Stream*, *Charles.* **4.** Banks of the River Wye between Tintern and Bigs-weir*; Pont y Saison; Shirenewton; Newchurch, *Shoolbred:* Redbrook*; near Penallt*, *Charles.*

Carex nigra (L.) Reichard   Common Sedge

Marshy places; common. All districts. Native.

Carex paniculata L.   Tussock Sedge

Marshy places and ditches; rare to locally frequent. Native.
2. Glebe Farm, Bedwas*. **3.** Pont-y-felin Wood, near Pontnewydd*, *Conway:* Llantarnam, *Hamilton:* Mescoed Mawr, Bettws; Michaelstone y Vedw*; Pwll Diwaelog, Castleton*; near Castell-y-bwch*; near Plas Machen. **4.** Rhyd-y-fedw, Shirenewton*; between Magor and Rogiet, *Shoolbred:* near The Virtuous Well, Trelleck*, *Shoolbred,!.*

CYPERACEAE

Carex otrubae Podp.   Fox Sedge
  C. vulpina auct.

Marshes, reens, and margins of ponds and streams; common. Districts 1, 3-5. Native.

(**Carex disticha** Huds. recorded by Clark is probably an error.)

Carex divulsa Stokes   Grey Sedge

Hedgebanks and woods; fairly common in Districts 1, 3 & 4, unrecorded from 2 & 5. Native.

× **otrobae**
4. Wood near Portskewett, with the parents*, *Marshall & Shoolbred.*

Carex leersii F. W. Schultz   Leer's Prickly Sedge

Roadside banks; rare. Native.
4. Near Tintern, *Bickham & E. F. Linton:* Usk Road, near Chepstow; between Caerwent and Llanvair Discoed*, *Shoolbred.*

Shoolbred's record from between the Wynd Cliff and Tintern is an error; his specimen is immature *C. spicata.*

Carex spicata Huds.   Prickly Sedge
  C. contigua Hoppe

Hedgebanks and railway banks; locally common. Native.
1. Common. **2.** Bedwas. **3.** Common. **4.** Common. **5.** Whitson; Goldcliff, *Hamilton.*

Carex muricata L.   Western Prickly Sedge
  C. pairaei F. W. Schultz

Roadside banks; rare. Native.
2. Ascent to the Sugar Loaf from Abergavenny, *Sandwith:* **3.** Near Rogerstone*. **4.** Lone Lane, Pentwyn, Penallt*, *Charles:* Wynd Cliff Wood.

Carex echinata Murr.   Star-headed Sedge

Wet moorland, acid marshes and bogs; common. Districts 1-4. Native.

Carex remota L.   Distant-spiked Sedge

Hedgebanks, ditches, shady river banks and woods; common. All districts. Native.

Carex curta Gooden.   White Sedge

Boggy places; rare. Native.
1. Redding's Enclosure*, *Charles:* **4.** Catbrook, near Tintern; Trelleck Bog*, *Shoolbred,!.*

CYPERACEAE

Carex ovalis Gooden.   Oval Sedge
Rough pastures, marshy places, wet heaths and bogs; common. All districts. Native.

Carex pulicaris L.   Flea Sedge
Acid marshes and boggy places; common. Districts 1-4. Native.

GRAMINEAE
PHRAGMITES Adans.
Phragmites australis (Cav.) Steud.   Reed
 P. communis Trin.
Swamps, marshes ditches and reens; very common in District 5, locally common elsewhere. All districts. Native.

MOLINIA Schrank
Molinia caerulea (L.) Moench   Purple Moor-grass
Moorland, wet heaths and boggy places; common. Districts 1-4. Native.

SIEGLINGIA Bernh.
Sieglingia decumbens (L.) Bernh.   Heath Grass
Heaths, pastures and railway banks; fairly common. Districts 1-4. Native.

GLYCERIA R. Br.
Glyceria fluitans (L.) R. Br.   Floating Meadow-grass
Ponds, streams and canals; common. All districts. Native.

× **plicata** (*G. × pedicellata* Townsend)
1. Pont-y-spig*, *R. Lewis.* **3.** The Park, Christchurch. **4.** Near Chepstow; Tintern; St. Arvans*; Llanvair Discoed, *Shoolbred:* Trelleck, *R. Lewis.* **5.** Near Pill-du Farm, St. Mellons*.

Glyceria plicata Fr.
By streams, ponds and canals; common. All districts. Native.

Glyceria declinata Bréb.
Marshes and muddy places; rare to frequent. Native.
3. Near Court Perrott, Llandegfedd*; Craig-y-ceiliog, Bettws*. **4.** Kilgwrrwg Bottom*; The Coombe, Shirenewton; Llanvair Discoed*, *Shoolbred:* Wentwood*; Trelleck Bog. **5.** Near Ffynnon Slwt, St. Mellons*.

GRAMINEAE

Glyceria maxima (Hartm.) Holmberg   Reed Meadow-grass

Pond, canal and reensides; locally common. Native.
2. Canal from Bassaleg to Crosskeys. **3.** Pontsampit Bridge, near Usk, *Clark:* canal, Malpas. **4.** Tintern*; Mathern*, *Shoolbred:* near Moynes Court, *T. G. Evans.* **5.** Newport*, *Conway:* Magor*, *Shoolbred:* Peterstone Wentloog*, *Harrison:* Rumney to St. Brides Wentlloog; near Llanwern.

FESTUCA L.

Festuca pratensis Huds.   Meadow Fescue

Meadows, pastures and grassy roadsides; common. All districts. Native.

Festuca arundinacea Schreb.   Tall Fescue

River banks, meadows and grassy roadsides; locally frequent. Native.
1. Between the Pentre and the Onen*, *Charles:* **4.** Tintern, *Ley, Shoolbred:* Penallt*, *Charles:* below Penallt Hill, Redbrook; banks of the River Wye, Tintern, *R. Lewis.* **5.** Sudbrook, *T. G. Evans.*

Festuca gigantea (L.) Vill.   Tall Brome

Woods and hedgebanks; common. Districts 1-4. Native.

Festuca altissima All.   Reed Fescue

Rocky woods, mainly on the limestone; rare. Native.
1. Grwyne Fawr Valley, *Ley:* Coleford Road near the county boundary*; Lilyrock Wood*; Lady Park Wood*; near Redding's Lodge*; Garth Wood*, *Charles.* **4.** Near Tintern, *Bicheno:* Wynd Cliff Wood*; *Ley, Bickham:* between Wynd Cliff and Tintern*, *Shoolbred,!.*

Festuca rubra L. ssp. **rubra**   Red Fescue

Pastures, roadsides, quarries, railway banks and mudflats; common. All districts. Native.

ssp. **commutata** Gaudin
4. 'Common', *Shoolbred.*

Festuca ovina L.   Sheep's Fescue

Heaths and dry grassland; common, chiefly in the upland districts. Districts 1-4. Native.

Festuca tenuifolia Sibth.   Narrow-leaved Fescue

Dry grassy banks; rare. Native.
1. 'Warfields', Staunton Road, Dixton Newton*, *R. Lewis.* **5.** Rumney*.

GRAMINEAE

×FESTULOLIUM Aschers. & Graebn.

×**Festulolium loliaceum** (Huds.) P. Fourn. (*Festuca pratensis* × *Lolium perenne*)

This sterile hybrid occurs occasionally with the parents.

LOLIUM L.

Lolium perenne L. Common Rye Grass

Very common in pastures, meadows, grassy roadsides and waste ground. All districts. Native but often sown in renovating grassland.

Lolium multiflorum Lam. Italian Rye Grass

Grassy roadsides and waste ground; a frequent escape from cultivation. Districts 1, 3 & 4.

Lolium temulentum L. Darnel

Waste places, rare. Casual.
1. Chippenham*, *Charles.* **4.** Earlswood Common, *Shoolbred.*

VULPIA C. C. Gmel.

Vulpia bromoides (L.) Gray Squirrel-tail Grass
Festuca bromoides L.

Quarries, claypits, walls, railway banks, colliery tips and dry gravelly places; frequent. All districts. Native.

Vulpia myuros (L.) C. C. Gmel. Rat's-tail Grass
Festuca myuros L.

Walls and wall bases, railway banks and waste ground; locally frequent. ?Native.
1. Penpergwm, *Reader:* The Kymin, Monmouth*; May Hill Station, Monmouth*; Skenfrith*; near Dingestow*; near Monmouth Troy Station*; near Dixton*, *Charles.* **3.** Llanbadoc, *Clark.* **4.** Near Tintern, *Hamilton, Shoolbred:* Barbadoes Hill, Tintern*, *Shoolbred:* Penallt*, *Charles:* Chepstow, *Shoolbred, T. G. Evans.* **5.** Newport Docks*.

PUCCINELLIA Parl.

Puccinellia maritima (Huds.) Parl. Sea Meadow-grass

Salt marshes and tidal banks of the rivers; common. Districts 4 & 5. Native.

Puccinellia distans (L.) Parl. Reflexed Meadow-grass

Salt marshes and tidal banks of rivers; rare. Native.
1. Near Wye Bridge, Monmouth*, *Charles.* **4.** Chepstow*; St. Pierre

GRAMINEAE

Pill*, *Shoolbred*. **5.** Between Sudbrook and Rogiet, *Shoolbred:* Rumney*, *Richards,!*.

Puccinellia rupestris (With.) Fernald & Weatherby
Procumbent Meadow-grass
Tidal banks of rivers and muddy places along sea walls; rare. Native.
4. Chepstow*, *Clark*, *Shoolbred*. **5.** Newport, *Clark:* Rumney*; St. Brides Wentlloog*.

CATAPODIUM Link

Catapodium rigidum (L.) C. E. Hubbard   Hard Fescue
Festuca rigida (L.) Rasp., non Roth; *Desmazeria rigida* (L.) Tutin
Walls, quarries, railway tracks and waste places; locally frequent to common. All districts. Native.

POA L.

Poa annua L.   Annual Meadow-grass
Grassy places, waste ground, cultivated ground, walls, etc.; very common. All districts. Native.

Poa nemoralis L.   Wood Meadow-grass
Woods, hedgebanks, shady cliffs and walls; common. Districts 1-4. Native.

Poa compressa L.   Flat-stalked Meadow-grass
Walls, quarries and rocky places; locally frequent. Native.
1. Abergavenny district, *White:* Llanthony Abbey, *Ley:* Monmouth, *Charles:* Llanfoist*. **3.** Usk, *Sandwith:* Caerleon. **4.** Troy House, *Riddelsdell*, *Charles:* Shirenewton*, *Hamilton*, *Shoolbred:* Chepstow*; Tintern*; Wynd Cliff; Piercefield; near Devauden*, *Shoolbred:* Ifton Quarries*; Magor*.

Poa pratensis L.   Smooth Meadow-grass
Grassy roadsides, waste ground, cultivated ground, walls and pastures; common. All districts. Native.

Poa angustifolia L.   Narrow-leaved Meadow-grass
Roadsides and waste ground; distribution uncertain. Native.
5. Newport Docks*.

Poa subcaerulea Sm.
Wall tops, mountain grassland and old colliery tips; ?frequent. Native.
2. Abersychan; Blorenge. **3.** Craigyceiliog, near Bettws*. **4.** Near Kemeys Inferior. **5.** Magor*, *Shoolbred:* Rumney.

216

Poa trivialis L.   Rough Meadow-grass

Meadows, pastures, grassy roadsides, cultivated and waste ground; very common and much more so in grassland than *Poa pratensis*. All districts. Native.

Poa palustris L.

Alien.

5. Newport Docks, 1954, *McClintock*.

Poa chaixii Vill.

Alien.

4. Lane bank by Pen-y-clawdd Church, 1946, *Sandwith*.

CATABROSA Beauv.

Catabrosa aquatica (L.) Beauv.   Whorl Grass

Muddy margins of ponds, rivers and reens; rare. Native.
4. Banks of the River Wye; St. Pierre Pond, *Shoolbred:* by the Virtuous Well, Trelleck*, *Charles*. **5.** Reens near the Severn, *Shoolbred:* near Rumney*; near Hendre-isaf, St. Mellons*.

DACTYLIS L.

Dactylis glomerata L.   Cocksfoot

Pastures, meadows, grassy roadsides, etc.; very common. All districts. Native.

CYNOSURUS L.

Cynosurus cristatus L.   Crested Dog's-tail

Very common in pastures, meadows and on grassy roadsides. All districts. Native.

BRIZA L.

Briza media L.   Quaking-grass

Pastures and meadows; common, especially on calcareous soils. All districts. Native.

MELICA L.

Melica uniflora Retz.   Wood Melic

Woods and shady hedgebanks; common. All districts. Native.

217

GRAMINEAE

Melica nutans L.   Mountain Melic

Woods, shady rocky banks and railway banks on the limestone; locally frequent. Native.

1. Lady Park Wood*; near Highmeadow Siding*; railway bank between Hadnock and The Bibblings, *Charles,!.* **4.** Piercefield Woods, *Lightfoot:* 2 miles west of Chepstow, *Hort:* about Tintern, *Motley, Gissing, Reader, Richards,!:* Wynd Cliff*, *E. Lees, Shoolbred, T. G. Evans,!:* Mounton*; near Pandy Mill, Itton, *Shoolbred:* Cuckoo Wood, Llandogo; Black Cliff, near Tintern.

First recorded by John Lightfoot in 1773.

BROMUS L.

Bromus erectus Huds.   Upright Brome

Pastures, meadows and grassy banks, chiefly on calcareous soils; locally common. Native.

1. Between Monmouth and the Fiddler's Elbow*; Leasebrook Lane; Beaulieu Wood*; Pont y Saison*; Hadnock*; Wyesham*, *Charles.* **4.** Portskewett*; St. Arvans*; *Shoolbred:* Upper Redbrook*; Penallt*, *Charles:* Shirenewton, *Rees.* **5.** Marshfield*.

Bromus ramosus Huds.   Hairy Brome
Zerna ramosa (Huds.) Lindm.

Woods, hedgebanks and thickets; common. All districts. Native.

Bromus benekenii (Lange) Trimen
B. ramosus var *benekenii* (Lange) Dr.

Woods and hedgebanks; rare. Native.
4. Piercefield Park; Wynd Cliff, *Ley:* between Wynd Cliff and Tintern*; Runston, *Shoolbred.*

Bromus sterilis L.   Barren Brome
Anisantha sterilis (L.) Nevski

Roadsides, waste and cultivated ground; common. All districts. Native.

Bromus madritensis L.   Wall Brome
Anisantha madritensis (L.) Nevski

Roadsides and wood borders; rare. Native.
4. Piercefield Woods, *Lightfoot*, 1773.

Bromus hordeaceus L.   Soft Brome
B. mollis L.

Meadows, pastures, grassy roadsides and waste ground; Native.

ssp. **hordeaceus**

Common. All districts.

ssp. **thominii** (Hardouin) Hylander

Roadsides and waste places; widespread and probably common. All districts except 3.

Bromus lepidus Holmberg

Roadsides and waste places, often sown in leys; frequent. Denizen.
1. Hadnock*; Llangua, between Monmouth Cap and Grosmont*; lane from Wonastow to King's Wood*; The Firs, near Monmouth*, *Charles:* Warfields, Staunton Road, near Monmouth; between Abergavenny and Crickhowell*, *R. Lewis.* **2.** Pant-glas Farm, Bedwas*. **3.** Near Coed Meredith, Henllys*. **4.** By Thicket Wood, near Rogiet*.

Bromus racemosus L.   Smooth Brome

Meadows and pastures; frequent. Native.
1. Wonastow*; Hadnock Farm*; *Charles:* Dingestow, *Sandwith.* **3.** By Park Wood, near Llanfihangel Llantarnam*. **4.** Piercefield Park*, *Ley, Shoolbred:* Rossfield Farm; Howick, *Shoolbred:* Penallt*; Duffield's Farm, Upper Redbrook*, *Charles:* between Wilcrick and Magor*. **5.** Peterstone Wentlloog*; Undy*; near Hendre Isaf, St. Mellons*.

Bromus commutatus Schrad.   Greater Smooth Brome

Meadows, pastures and grassy roadsides; widespread but not common. Native.
1. Hadnock Farm, near Monmouth*; Wonastow*; Wyesham, *Charles:* **3.** Llangybi, *Clark:* near Castell-y-bwch, Henllys*. **4.** Near Chepstow; Portskewett; Ifton Hill; Tintern, *Shoolbred.* **5.** Rumney*.

Bromus secalinus L.   Rye Brome

Waste places; rare. Casual.
1. Ysgubor Newydd, Llanfrechfa Lower*, *Conway.* **5.** Coast Road, near Rumney*.

BRACHYPODIUM Beauv.

Brachypodium sylvaticum (Huds.) Beauv.   Wood False Brome

Woods, hedgebanks and rough pastures; common. All districts. Native.

Brachypodium pinnatum (L.) Beauv.   Heath False Brome

Pastures and woodland rides on calcareous soils; rare. Native.
1. Redding's Enclosure*; Lady Park Wood*, *Charles,!.* **4.** Near Penallt*, *Charles.*

219

GRAMINEAE

AGROPYRON Gaertn.

Agropyron caninum (L.) Beauv.   Tufted Couch-grass

Woods, shady hedgebanks and wooded river banks; locally frequent. Native.

1. Llanthony Valley, *Ley:* banks of the River Wye and woods about Monmouth*, *Charles.* **3.** Llanrumney. **4.** Piercefield Woods, *Lightfoot:* Chepstow; Wynd Cliff*; Tintern; Mounton; near Crossway Green; The Coombe, Shirenewton, *Shoolbred:* Black Cliff, near Tintern*.

First recorded by John Lightfoot in 1773.

Agropyron repens (L.) Beauv.   Couch-grass, Twitch

Cultivated ground, waste ground and roadsides; very common. All districts. Native.

Agropyron pungens (Pers.) Roem. & Schult.   Sharp Couch-grass

Mud flats and tidal banks of rivers; locally common. Native.
4. Chepstow, *Shoolbred.* **5.** Newport*, *Conway:* between Sudbrook and Rogiet, *Shoolbred:* Portskewett, *Hamilton:* Denny Island, *Matthews:* Magor Pill*.

HORDEUM L.

Hordeum secalinum Schreb.   Meadow Barley

Moist pastures and meadows, especially near the sea; locally common. Native.
3. Caerleon. **4.** Chepstow; Piercefield*; Tintern*, *Shoolbred:* near Mitchel Troy*, *Charles.* **5.** Common and often abundant along Severn.

Hordeum murinum L.   Wall Barley

Roadsides, waste ground and wall bases; frequent to locally common. Native.
1. Monmouth, *Charles:* Llanfoist. **2.** Griffithstown. **3.** Between Newport and Bettws Road. **4.** Chepstow, *Clark:* Portskewett; Magor, *Shoolbred.* **5.** Newport*†, *Conway, Clark:* Sudbrook*, *Shoolbred:* Rumney*; Peterstone Wentlloog*.

Hordeum marinum L.   Sea Barley

Salt marshes; rare. Native.
4. Chepstow, *Watkins.* **5.** Between Sudbrook and Rogiet*, *Shoolbred:* Peterstone Wentlloog*.

GRAMINEAE

HORDELYMUS (Jessen) Harz

Hordelymus europaeus (L.) Harz   Wood Barley

Woods on the limestone; rare. Native.

1. Near Highmeadow Siding; Lady Park Wood*; Redding's Enclosure*, *Charles,!.* **4.** Piercefield Woods, *Lightfoot:* Wynd Cliff*, *Purchas, C. E. Salmon.*

First recorded by John Lightfoot in 1773.

KOELERIA Pers.

Koeleria cristata (L.) Pers.   Slender Hair-grass

Limestone pastures and quarries, rare. Native.

4. Near Portskewett*, *Marshall & Shoolbred:* Ifton Quarries*.

Hamilton's record from the Rhymney Valley is probably an error.

TRISETUM Pers.

Trisetum flavescens (L.) Beauv.   Yellow Oat-grass

Pastures and grassy roadsides; common in Districts 1, 3 and 4, rare or absent from 2 and 5. Native.

AVENA L.

Avena fatua L.   Wild Oat

Cultivated fields and waste places; locally common. Colonist.

4. 'Common in the Chepstow district', *Shoolbred:* Llanvair Discoed, *Lannon.* **5.** Undy*.

Avena strigosa Schreb.

Cultivated fields and roadsides; rare. Colonist or casual.

1. 'Warfields,' Dixton Newton Parish, *R. Lewis**. **4.** Tintern, *Gambier-Parry.* **5.** Rumney*, *Miss Vachell.*

Avena sativa L.   Oat

Roadsides and waste places; a frequent casual.

HELICTOTRICHON Bess.

Helictotrichon pratense (L.) Pilg.   Meadow Oat-grass

Recorded in Watson's *Topographical Botany*, ed. 1, 1874 on the authority of Conway.

Helictotrichon pubescens (Huds.) Pilg.   Downy Oat-grass

Pastures, meadows, grassy roadsides and railway banks; common on the calcareous soils of Districts 1 and 2. Native.

221

GRAMINEAE

ARRHENATHERUM Beauv.

Arrhenatherum elatius (L.) Beauv. ex J. & C. Presl   False Oat-grass

Hedgebanks, wood borders, pastures and waste places; very common. All districts. Native.

HOLCUS L.

Holcus lanatus L.   Yorkshire Fog

Meadows, pastures and grassy roadsides; very common. All districts. Native.

Holcus mollis L.   Creeping Soft-grass

Woods, hedgebanks and heaths; locally frequent to common. Districts 1-4 Native.

DESCHAMPSIA Beauv.

Deschampsia caespitosa (L.) Beauv.   Tufted Hair-grass

Marshes, ill-drained pastures and damp open woods; very common. All districts. Native.

Deschampsia flexuosa (L.) Trin.   Heath Hair-grass

Heaths, moorlands, woods and turfy wall tops; common, especially on the higher ground. Districts 1-4. Native.

AIRA L.

Aira praecox L.   Early Hair-grass

Heaths, dry banks, wall tops, dry pastures and colliery tips; very common in District 2, locally common in 1, 3 and 4. Native.

Aira caryophyllea L.   Silvery Hair-grass

Dry pastures, walls, quarries and colliery tips; frequent to locally common. All districts. Native.

CALAMAGROSTIS Adans.

Calamagrostis epigejos (L.) Roth   Wood Small-reed

Woods, dry banks and quarries on the limestone; locally frequent. Native. **1.** Wyesham*, *Charles:* **4.** Usk Road, near Chepstow*; The Barnetts; Portskewett; Wynd Cliff; near St. Brides Netherwent, etc., *Shoolbred:* near Chepstow, *Hamilton.*

GRAMINEAE

AGROSTIS L.

Agrostis canina L.   Brown Bent

Bogs, acid marshes, wet heaths and damp rough pastures; rather rare. Native.
1. Near Pont-y-spig*. **3.** Near Pontneweydd*, *Conway*. **4.** Trelleck Bog*, *Watkins, Shoolbred,!:* Mounton; Portskewett; Earlswood Common, *Shoolbred:* Whitelye Common*.

Agrostis tenuis Sibth.   Fine Bent

Meadows, pastures, heaths, grassy roadsides and woodland rides; very common. All districts. Native.

Agrostis gigantea Roth.   Black Bent

Cultivated and waste ground; rare. Colonist.
1. Llanddewi Ysgyryd, *Ley:* Monmouth*, *Charles*. **4.** Redbrook*, *Charles*.

Shoolbred's record from Whitebrook is erroneous.

Agrostis stolonifera L.   White Bent

Pastures, grassy roadsides, ditches and waste ground; common. All districts. Native.

var. **palustris** (Huds.) Farw. is common along the reens by the Severn.

GASTRIDIUM Beauv.

Gastridium ventricosum (Gouan) Schinz & Thell.   Nit Grass

Waste ground; rare. Alien.
4. Chepstow, *Watkins*. **5.** Newport Docks, *Clark*.

PHLEUM L.

Phleum bertolonii DC.   Cat's-tail
  P. nodosum auct.

Dry pastures and roadsides; frequent. All districts. Native.

Phleum pratense L.   Timothy Grass

Pastures, meadows and grassy roadsides; common. All districts. Often sown for hay grass and probably only native in moist lowland grassland.

ALOPECURUS L.

Alopecurus myosuroides Huds.   Field Fox-tail

Cultivated ground and waste places; rare to locally frequent. Colonist.
1. Near Rockfield Road, Monmouth*; Hendre*; St. Maughans*; near Chippenham*; Rockfield*; Wonastow*; Kymin Hill*, *Charles*. **3.** Gwernesney, *Rees*. **4.** Chepstow district, *Shoolbred*. **5.** Dyffryn, *Hamilton*.

GRAMINEAE

Alopecurus pratensis L.   Meadow Fox-tail

Meadows, pastures and grassy roadsides; very common. All districts. Native.

Alopecurus geniculatus L.   Knee-bent Fox-tail

Marshes and margins of ponds and reens; common. All districts. Native.

Alopecurus aequalis Sobol.   Orange Fox-tail

Very rare. Native.
1. Margin of a pond near Penpergwm Station, 1929*, *Rickards.*

Alopecurus bulbosus Gouan   Bulbous Fox-tail

Salt marshes and fields frequently inundated by the tide; locally common. Native.
3. By the River Usk, Malpas. **4.** 'Marshes by the Severn side going the footway from the New Passage House to Chepstow . . . plentifully', *Lightfoot:* by the Wye, Chepstow, *Shoolbred.* **5.** Rumney Marshes*, *Lightfoot,!:* St. Pierre Pill†, *Clark:* Newport Marshes, *Hamilton:* below Mathern*, *Shoolbred:* Collister Pill, near Undy*.

First recorded by *John Lightfoot* in 1773.

MILIUM L.

Milium effusum L.   Wood Millet

Woods; common in Districts 1-4, unrecorded from 2 & 5. Native.

ANTHOXANTHUM L.

Anthoxanthum odoratum L.   Sweet Vernal Grass

Meadows, pastures, roadsides and woods; common. All districts. Native.

PHALARIS L.

Phalaris arundinacea L.   Ribbon Grass

Marshes and by rivers, streams and reens; common. All districts. Native.

Phalaris canariensis L.   Canary Grass

An occasional casual on waste ground. Districts 1, 3 & 4.

PARAPHOLIS C. E. Hubbard

Parapholis strigosa (Dumort.) C. E. Hubbard   Hard Grass

Salt marshes; common along the Severn and the tidal banks of the Rivers Wye and Usk. Districts 4 & 5. Native.

GRAMINEAE

NARDUS L.

Nardus stricta L.   Mat Grass

Heaths, moors and dry pastures on acid soils; very common in the upland areas, locally common elsewhere. Districts 1-4. Native.

SPARTINA Schreb.

Spartina anglica C. E. Hubbard.   Rice Grass

Salt marshes and mud flats; established in abundance in many places along the Severn. Denizen.

Spartina anglica is derived from the sterile primary hybrid *S. alternifolia* ×*maritima* by doubling of the chromosomes.

First recorded by Miss E. Vachell from near Portskewett in 1935.

CYNODON Rich.

Cynodon dactylon (L.) Pers.   Bermuda Grass

Established alien.

2. Colliery tip, South Celynen Colliery, Abercarn*, 1959, *Palmer*.

PANICUM L.

Panicum miliaceum L.   Millet

Casual.

3. Among root crops, Bassaleg*.

ECHINOCHLOA Beauv.

Echinochloa crus-galli (L.) Beauv.   Cockspur

Cultivated ground. Casual.

1. Fiddler's Elbow, near Monmouth*, *R. Lewis*.

DIGITARIA P. C. Fabr.

Digitaria ischaemum (Shreb.) Muhl.   Red Millet

Casual.

5. Newport Docks, *Hamilton*.

SETARIA Beauv.

Setaria viridis (L.) Beauv.   Green Bristle Grass

Casual.

1. Garden weed, The Kymin, Monmouth*, *Charles*. 4. Chepstow, *Ley*.

BECKMANNIA Host

Beckmannia erucaeformis (L.) Host

Casual.

1. Railway between Monmouth and Dixton*, *Charles*.

INDEX

OF THE FAMILIES AND GENERA AND OF THE
SUBSTANTIVE ENGLISH NAMES

Acer, 87
ACERACEAE, 87
Achillea, 179
Acinos, 160
Aconitum, 52
Adder's Tongue, 50
Adonis, 55
Adoxa, 170
ADOXACEAE, 170
Aegopodium, 124
Aesculus, 87
Aethusa, 125
Agrimonia, 104
Agrimony, 104
Agropyron, 220
Agrostemma, 73
Agrostis, 223
Aira, 222
Ajuga, 164
Alchemilla, 104
Alder, 133
Alexanders, 122
Alisma, 191
ALISMATACEAE, 191
Alkanet, 145
Alliaria, 67
Allium, 199
Alnus, 133
Alopecurus, 223
Althaea, 82
Alyssum, 64
AMARANTHACEAE, 78
Amaranthus, 78
AMARYLLIDACEAE, 199
Ambrosia, 173
American Cudweed, 177
Amsinckia, 145
Anacamptis, 205
Anagallis, 141
Anaphalis, 177
Anchusa, 145
Anemone, 52
Angelica, 126
Antennaria, 177
Anthemis, 178
Anthoxanthum, 224
Anthriscus, 120
Anthyllis, 93
Antirrhinum, 150
Aphanes, 105
Apium, 122
APOCYNACEAE, 142

Apple, 110
AQUIFOLIACEAE, 87
Aquilegia, 55
Arabidopsis, 68
Arabis, 65
ARACEAE, 205
ARALIACEAE, 119
Arctium, 181
Arenaria, 76
Armeria, 139
Armoracia, 64
Arrhenatherum, 222
Arrow-grass, 192
Arrowhead, 191
Artemisia, 180
Arum, 205
Ash, 142
Asparagus, 195
Aspen, 135
Asperula, 167
ASPIDIACEAE, 48
ASPLENIACEAE, 46
Asplenium, 46
Aster, 177
Astrantia, 120
Astragalus, 94
ATHYRIACEAE, 47
Athyrium, 47
Atriplex, 79
Atropa, 149
Avena, 221
Avens, 103
Axyris, 80
Azolla, 50
AZOLLACEAE, 50

Baldellia, 191
Ballota, 162
Balm, 161
Balsam, 86
BALSAMINACEAE, 86
Barbarea, 65
Barberry, 56
Barley, 220
Barren Strawberry, **102**
Bartsia, 156
Basil Thyme, 160
Beakrush, 208
Beckmannia, 225
Bedstraw, 167
Beech, 134
Beech Fern, 49

Beet, 79
Bellflower, 165
Bellis, 178
Bent, 223
BERBERIDACEAE, 56
Berberis, 56
Bermuda Grass, 225
Berteroa, 64
Berula, 124
Beta, 79
Betonica, 161
Betony, 161
Betula, 133
BETULACEAE, 133
Bidens, 172
Bilberry, 138
Bindweed, 148
Bindweed, Black, 129
Birch, 133
Birdsfoot, 94
Birdsfoot Trefoil, 93
Birds-nest, 139
Bistort, 128
Bitter-cress, 64
Blackberry, 98
Blackstonia, 143
Blackthorn, 107
Bladder-fern, 47
Bladderwort, 157
BLECHNACEAE, 46
Blachnum, 46
Bluebell, 196
Bog Asphodel, 194
Bogbean, 144
BORAGINACEAE, 144
Botrychium, 50
Brachypodium, 219
Bracken, 46
Bramble, 97, 98
Brassica, 59
Bridewort, 97
Bristle Grass, 225
Briza, 217
Brome, 214, 218
Bromus, 218
Brooklime, 153
Brookweed, 142
Broom, 89
Broomrape, 156
Bryonia, 126
Bryony, Black, 201
Bryony, White, 126
Buckler-fern, 48
Buckthorn, 88
Buckwheat, 130
Buddleja, 142
BUDDLEJACEAE, 142
Bugle, 164
Bugloss, 146, 147

Bullace, 107
Bulrush, 207
Bupleurum, 122
Burdock, 181
Bur-Marigold, 172
Burnet, 105
Burnet Saxifrage, 124
Bur-reed, 206
Butcher's Broom, 195
BUTOMACEAE, 191
Butomus, 191
Butterbur, 175
Buttercup, 52
Butterwort, 157

Cabbage, 59
Calamagrostis, 222
Calamint, 160
Calamintha, 160
CALLITRICHACEAE, 118
Callitriche, 118
Calluna, 137
Caltha, 51
Calystegia, 148
Camelina, 68
Campanula, 166
CAMPANULACEAE, 165
Campion, 72
Canadian Waterweed, 192
Canary Grass, 224
Candytuft, 62
CANNABIACEAE, 132
CAPRIFOLIACEAE, 169
Capsella, 63
Caraway, 123
Cardamine, 64
Cardaria, 62
Carduus, 182
Carex, 208
Carlina, 181
Carpinus, 133
Carrot, 126
Carum, 123
CARYOPHYLLACEAE, 72
Castanea, 134
Castor-oil Plant, 128
Catabrosa, 217
Catapodium, 216
Catchfly, 72
Catmint, 163
Cat's-ear, 185
Cat's-tail, 223
Caucalis, 121
Celandine, Greater, 58
Celandine, Lesser, 55
CELASTRACEAE, 87
Celery, 122
Centaurea, 183
Centaurium, 143

228

Centaury, 143
Centranthus, 171
Cephalanthera,, 201
Cerastium, 74
CERATOPHYLLACEAE, 57
Ceratophyllum, 57
Ceterach, 47
Chaenorrhinum, 151
Chaerophyllum, 120
Chamaemelum, 179
Chamomile, 178
Charlock, 60
Cheiranthus, 67
Chelidonium, 58
CHENOPODIACEAE, 78
Chenopodium, 78
Cherry, 107
Cherry Laurel, 108
Chervil, 120
Chickweed, 74
Chicory, 184
Chives, 199
Chrysanthemum, 180
Chrysosplenium, 112
Cichorium, 184
Cicuta, 123
Cinquefoil, 102
Cirsium, 182
Circaea, 117
CISTACEAE, 72
Claytonia, 78
Clematis, 52
Clinopodium, 161
Clover, 91
Clubmoss, 45
Clubrush, 207
Cochlearia, 63
Cocksfoot, 217
Cockspur, 225
Coeloglossum, 202
Colchicum, 196
Coltsfoot, 174
Columbine, 55
Comfrey, 145
COMPOSITAE, 172
Conium, 122
Conopodium, 123
Conringia, 61
Convallaria, 195
CONVOLVULACEAE, 148
Convolvulus, 148
Conyza, 178
Coriander, 121
Coriandrum, 121
CORNACEAE, 119
Corn Caraway, 123
Corn Cockle, 73
Cornflower, 184
Corn Marigold, 180

Corn-salad, 170
Corn Spurrey, 77
Coronilla, 94
Coronopus, 62
Cornus, 119
Corydalis, 58
CORYLACEAE, 133
Corylus, 134
Cotoneaster, 108
Cotton-grass, 206
Cotyledon, 111
Couch-grass, 220
Cowberry, 138
Cowslip, 140
Cow-wheat, 155
Cranberry, 138
Cranesbill, 83
CRASSULACEAE, 110
Crataegus, 108
Creeping Jenny, 140
Crepis, 190
Cress, 62
Crithmum, 124
Crocosmia, 201
Crocus, 201
Crosswort, 167
Crowberry, 139
Crowfoot, 53
Crown Vetch, 94
Cruciata, 167
CRUCIFERAE, 59
Cuckoo Pint, 205
CURCUBITACEAE, 126
Cudweed, 176
Currant, 113
Cuscuta, 148
Cyclamen, 140
Cymbalaria, 151
Cynodon, 225
Cynoglossum, 144
Cynosurus, 217
CYPERACEAE, 206
Cystopteris, 47
Cytisus, 89

Dactylis, 217
Dactylorhiza, 204
Daffodil, 200
Daisy, 178
Dame's Violet, 67
Dandelion, 190
Danewort, 169
Daphne, 114
Darnel, 215
Datura, 149
Daucus, 126
Deadnettle, 162
Deerhair Sedge, 207

Delphinium, 52
DENNSTAEDTIACEAE, 46
Deschampsia, 222
Descurainia, 68
Desmazeria, 216
Dewberry, 97
Dianthus, 73
Digitalis, 152
Digitaria, 225
DIOSCOREACEAE, 201
Diplotaxis, 60
DIPSACACEAE, 172
Dipsacus, 172
Dittander, 61
Dock, 130
Dodder, 148
Dog Daisy, 180
Dog's-tail 217
Dogwood, 119
Doronicum, 174
Dropwort, 97
Drosera, 114
DROSERACEAE, 114
Dryopteris, 48
Duckweed, 205

Echinochloa, 225
Echium, 147
Elder, 169
Eleocharis, 207
Elecampane, 175
Elodea, 192
Elm, 132
EMPETRACEAE, 139
Empetrum, 139
Enchanter's Nightshade, 117
Endymion, 196
Epilobium, 115
Epipactis, 201
EQUISETACEAE, 45
Equisetum, 45
Erica, 137
ERICACEAE, 137
Erigeron, 178
Erinus, 152
Eriophorum, 206
Erodium, 85
Erophila, 64
Eryngium, 120
Erysimum, 67
Euonymus, 87
Eupatorium, 178
Euphrasia, 155
Euphorbia, 127
EUPHORBIACEAE, 127
Evening Primrose, 117
Eyebright, 155

FAGACEAE, 134

Fagopyrum, 130
Fagus, 134
False Brome, 219
False Oat-grass, 222
Fennel, 125
Fenugreek, 91
Fescue, 214
Festuca, 214
Festulolium, 215
Feverfew, 180
Field Madder, 167
Figwort, 151
Filago, 176
Filipendula, 97
Flag, 200
Flax, 82
Fleabane, 176, 178
Fleece-flower, 130
Flixweed, 68
Flowering Rush, 191
Fox-tail, 223
Fluellen, 151
Foeniculum, 125
Fool's Parsley, 125
Forget-me-not, 146
Foxglove, 152
Fragaria, 103
Frangula, 88
Fraxinus, 142
Frog-bit, 192
Fumaria, 58
FUMARIACEAE, 58
Fumitory, 58
Furze, 89

Galanthus, 199
Galega, 93
Galeobdolon, 162
Galeopsis, 163
Galium, 167
Garlic, 199
Garlick Mustard, 67
Gastridium, 223
Gean, 107
Gentiana, 143
GENTIANACEAE, 143
Gentianella, 143
Gentian, 143
Genista, 88
GERANIACEAE, 83
Geranium, 83
Geum, 103
Gipsywort, 160
Gladdon, 200
Glasswort, 80
Glaucium, 58
Glaux, 141
Glechoma, 163

Globe Flower, 51
Glyceria, 213
Glycine, 96
Gnaphalium, 176
Goatsbeard, 186
Goat's Rue, 93
Goldenrod, 177
Gold of Pleasure, 68
Golden Saxifrage, 112
Goldilocks, 53
Good King Henry, 78
Gooseberry, 113
Goosefoot, 78
Goosegrass, 168
Gorse, 89
Goutweed, 124
GRAMINEAE, 213
Greenweed, 88
Groenlandia, 194
Gromwell, 147
GROSSULARIACEAE, 113
Ground Ivy, 163
Groundsel, 174
Guelder Rose, 169
Gunnera, 118
Gymnadenia, 203
Gymnocarpium, 49

Habenaria, 202
Hair-grass, 221, 222
HALORAGACEAE, 117
Hard Fern, 46
Hard Fescue, 216
Hard Grass, 224
Harebell, 166
Hare's-ear, 122
Hart's-tongue Fern, 46
Hawkbit, 185
Hawksbeard, 190
Hawkweed, 187
Hawthorn, 108
Hazel, 134
Heather, 137
Heath-grass, 213
Hedera, 119
Hedge Mustard, 67
Hedge Parsley, 121
Helianthemum, 72
Helictotrichon, 221
Hellebore, 51
Helleborine, 201
Helleborus, 51
Hemlock, 122
Hemp Agrimony, 178
Hempnettle, 163
Henbane, 149
Henbit, 162
Heracleum, 126
Herb Bennet, 103

Herb Paris, 196
Herb Robert, 85
Hesperis, 67
Hieracium, 187
Himalayan Honeysuckle, 170
HIPPOCASTANACEAE, 87
Hippocrepis, 94
HIPPURIDACEAE, 118
Hippuris, 118
Hirschfeldia, 60
Hogweed, 126
Holcus, 222
Holly, 87
Honeysuckle, 170
Honkenya, 76
Hop, 132
Hordelymus, 221
Hordeum, 220
Horehound, Black, 162
Horehound, White, 164
Hornbeam, 133
Horned Pondweed, 194
Hornwort, 57
Horse Chestnut, 87
Horse Radish, 64
Horsetail, 45
Horse-shoe Vetch, 94
Hound's-tongue, 144
Houseleek, 111
Humulus, 132
Hydrocharis, 192
HYDROCHARITACEAE, 192
Hydrocotyle, 119
Hyoscyamus, 149
HYPERICACEAE, 70
Hypericum, 70
Hypochaeris, 185

Iberis, 62
Ilex, 87
ILLECEBRACEAE, 77
Impatiens, 86
Inula, 175
IRIDACEAE, 200
Iris, 200
Isolepis, 207
Ivy, 119

Jacob's Ladder, 144
Jasione, 167
JUGLANDACEAE, 133
Juglans, 133
JUNCACEAE, 196
JUNCAGINACEAE, 192
Juncus, 196

Kickxia, 151
Kidney Vetch, 93
Knapweed, 183

Knautia, 172
Knawel, 77
Knotweed, 128
Koeleria, 221

LABIATAE, 158
Lactuca, 186
Lady-fern, 47
Lady's-mantle, 104
Lady's-smock, 64
Lady's-tresses, 202
Lagarosiphon, 192
Lamiastrum, 162
Lamium, 162
Lappula, 144
Lapsana, 185
Larkspur, 52
Lathraea, 156
Lathyrus, 96
Lavatera, 82
LEGUMINOSAE, 88
Lemna, 205
LEMNACEAE, 205
LENTIBULARIACEAE, 157
Leonurus, 163
Leopard's Bane, 174
Leontodon, 185
Lepidium, 61
Lettuce, 186
Leucanthemum, 180
Leucojum, 199
Leycesteria, 170
Ligustrum, 142
Lilac, 142
LILIACEAE, 194
Lilium, 195
Lily, 195
Lily of the Valley, 195
Lime, 81
Limestone Polypody, 49
Limonium, 139
LINACEAE, 82
Linaria, 150
Ling, 137
Linum, 82
Listera, 202
Lithospermum, 147
Littorella, 165
Lobularia, 64
Lolium, 215
Lonicera, 170
Loosestrife, Purple, 114
Loosestrife, Yellow, 141
LORANTHACEAE, 119
Lotus, 93
Lousewort, 154
Lucerne, 90
Lungwort, 146

Luzula, 198
Lychnis, 73
Lycium, 148
LYCOPODIACEAE, 45
Lycopodium, 45
Lycopsis, 146
Lycopus, 160
Lysimachia, 140
LYTHRACEAE, 114
Lythrum, 114

Madder, 169
Mahonia, 56
Male-fern, 48
Mallow, 81
Malus, 110
Malva, 81
MALVACEAE, 81
Maple, 87
Marestail, 118
Marjorum, 160
Marrubium, 164
Marsh Fern, 49
Marsh Mallow, 82
Marsh Marigold, 51
Marsh Pennywort, 119
Marshwort, 122
Mat Grass, 225
Matricaria, 179
Mayweed, 178, 179
Meadow-grass, 213, 215
Meadow Rue, 56
Meadow Saffron, 196
Meadowsweet, 97
Meconopsis, 57
Medicago, 90
Medick, 90
Melampyrum, 155
Melic, 217
Melica, 217
Melilot, 90
Melilotus, 90
Melissa, 161
Mentha, 158
Menyanthes, 144
Mercurialis, 127
Mercury, 127
Mezerion, 114
Mignonette, 68
Milium, 224
Milkwort, 70
Millet, 224, 225
Mimulus, 152
Mint, 158
Minuartia, 76
Misopates, 150
Mistletoe, 119
Moehringia, 76
Monkshood, 52

232

Molinia, 213
Moneywort, 140
Monkey-flower, 152
Monotropa, 139
MONOTROPACEAE, 139
Montbretia, 201
Montia, 77
Moonwort, 50
Moor-grass, 213
Moschatel, 170
Motherwort, 163
Mountain Ash, 108
Mountain Everlasting, 177
Mountain Fern, 48
Mugwort, 180
Mullein, 149
Musk, 152
Mustard, 59
Mycelis, 186
Myosotis, 146
Myosoton, 74
Myriophyllum, 117
Myrrhis, 121

Narcissus, 200
Nardus, 225
Narthecium, 194
Nasturtium, 66
Neottia, 202
Nepeta, 163
Nettle, 132
Nightshade, 149
Nipplewort, 185
Nit Grass, 223
Nuphar, 56
Nymphaea, 56
NYMPHAEACEAE, 56

Oak, 134
Oak Fern, 49
Oat, 221
Oat-grass, 221
Odontites, 156
Oenanthe, 124
Oenothera, 117
OLEACEAE, 142
ONAGRACEAE, 115
Onobrychis, 94
Ononis, 89
Onopordon, 183
OPHIOGLOSSACEAE, 50
Ophioglossum, 50
Ophrys, 204
Orache, 79
Orchid, 202
ORCHIDACEAE, 201
Orchis, 204
Oregon Grape, 56

Origanum, 160
Ornithogalum, 196
Ornithopus, 94
OROBANCHACEAE, 156
Orobanche, 156
Orpine, 110
Orthilia, 139
Osier, 136
Osmunda, 46
OSMUNDACEAE, 46
OXALIDACEAE, 86
Oxalis, 86
Ox-tongue, 185
Oxycoccus, 138

Panicum, 225
Pansy, 70
Papaver, 57
PAPAVERACEAE, 57
Parapholis, 224
Parietaria, 131
Paris, 196
Parsley, 123
Parsley Piert, 105
Parsnip, 126
Pastinaca, 126
Pea, 96
Pear, 109
Pearlwort, 75
Pedicularis, 154
Pellitory-of-the-wall, 131
Penny-cress, 63
Pennyroyal, 158
Pentaglottis, 145
Peplis, 114
Pepper-wort, 61
Periwinkle, 142
Persicaria, 129
Petasites, 175
Petroselinum, 123
Peucedanum, 126
Phalaris, 224
Phaseolus, 97
Pheasant's Eye, 55
Phleum, 223
Phragmites, 213
Phyllitis, 46
Phytolacca, 81
PHYTOLACCACEAE, 81
Picris, 185
Pignut, 123
Pimpernel, 141
Pimpinella, 124
PINACEAE, 51
Pine, 51
Pinguicula, 157
Pink, 73
Pinus, 51
PLANTAGINACEAE, 165

Plantago, 165
Plantain, 165
Platanthera, 203
Ploughman's Spikenard, 175
Plum, 107
PLUMBAGINACEAE, 139
Poa, 216
Poke Berry, 81
POLEMONIACEAE, 144
Polemonium, 144
Polygala, 70
POLYGALACEAE, 70
POLYGONACEAE, 128
Polygonatum, 195
Polygonum, 128
POLYPODIACEAE, 49
Polypodium, 49
Polypody, 49
Polystichum, 48
Pondweed, 192
Poplar, 135
Poppy, 57
Populus, 135
PORTULACACEAE, 77
Potamogeton, 192
POTAMOGETONACEAE, 192
Potentilla, 102
Poterium, 105
Primrose, 140
Primula, 140
PRIMULACEAE, 140
Privet, 142
Prunella, 161
Prunus, 107
Pseudorchis, 203
Pteridium, 46
Puccinellia, 215
Pulicaria, 176
Pulmonaria, 146
Pyrola, 138
PYROLACEAE, 138
Pyrus, 109

Quaking Grass, 217
Quercus, 134

Radish, 61
Ragged Robin, 73
Ragwort, 173
Ramischia, 139
Ramsons, 199
RANUNCULACEAE, 51
Ranunculus, 52
Rape, 59
Raphanus, 61
Raspberry, 97
Rat's-tail Grass, 215
Red-rattle, 154
Reed, 213

Reed Mace, 206
Reseda, 68
RESEDACEAE, 68
Restharrow, 89
RHAMNACEAE, 88
Rhamnus, 88
Rhinanthus, 155
Rhynchosinapis, 60
Rhynchospora, 208
Ribbon Grass, 224
Ribes, 113
Rice Grass, 225
Ricinus, 128
Rock-Cress, 65
Rocket, 60, 67
Rockrose, 72
Rorippa, 66
Rosa, 105
ROSACEAE, 97
Rose, 105
Royal Fern, 46
Rubia, 169
RUBIACEAE, 167
Rubus, 97
Rumex, 130
Ruppia, 194
RUPPIACEAE, 194
Ruscus, 195
Rush, 196
Rusty-back Fern, 47
Rye Grass, 215

Sage, 161
Sagina, 75
Sagittaria, 191
Sainfoin, 94
Saffron, 201
SALICACEAE, 135
Salicornia, 80
Salix, 135
Sallow, 136
Salsify, 186
Salvia, 161
Sambucus, 169
Samolus, 142
Samphire, 124
Sand Spurrey, 77
Sandwort, 76
Sanguisorba, 105
Sanicula, 120
Saponaria, 73
Sarothamnus, 89
Saw-wort, 184
Saxifraga, 112
SAXIFRAGACEAE, 112
Saxifrage, 112
Scabiosa, 172
Scabious, 172

234

Scandix, 120
Scirpus, 207
Scleranthus, 77
Scorpion-grass, 147
Scots Pine, 51
Scrophularia, 151
SCROPHULARIACEAE, 149
Scurvy-grass, 63
Scutellaria, 164
Sea-blite, 80
Sea Lavender, 139
Sea Milkwort, 141
Sea Purslane, 76
Sea Spurrey, 77
Sedge, 208
Sedum, 110
Selfheal, 161
Sempervivum, 111
Senebiera, 62
Senecio, 173
Serratula, 184
Service-tree, 109
Setaria, 225
Sheep's-bit Scabious, 167
Shepherd's Needle, 120
Shepherd's Purse, 63
Sherardia, 167
Shield-fern, 48
Shoreweed, 165
Sieglingia, 213
Silaum, 125
Silene, 72
Silverweed, 102
Silybum, 183
Sinapis, 60
Sison, 123
Sisymbrium, 67
Sium, 124
Skull-cap, 164
Small-reed, 222
Smyrnium, 122
Snapdragon, 150
Sneeze-wort, 179
Snowberry, 169
Snowdrop, 199
Snowflake, 199
Soapwort, 73
Soft-grass, 222
SOLANACEAE, 148
Solanum, 149
Solidago, 177
Solomon's Seal, 195
Sonchus, 187
Sorbus, 108
Sorrel, 130
Sow-thistle, 187
Soya-bean, 96
SPARGANIACEAE, 206
Sparganium, 206

Spartina, 225
Spearwort, 53
Speedwell, 153
Spergula, 77
Spergularia, 77
Spindle-tree, 87
Spiraea, 97
Spiranthes, 202
Spleenwort, 46
Spurge, 127
Spurge Laurel, 115
Squirrel-tail Grass, 215
Stachys, 162
Star-of-Bethlehem, 196
Stellaria, 74
Stitchwort, 75
St. John's-wort, 70
Stonecrop, 111
Stone Parsley, 123
Storksbill, 85
Strawberry, 103
Suaeda, 80
Succisa, 172
Sulphur-wort, 125
Sundew, 114
Sweet Alyssum, 64
Sweet Chestnut, 134
Sweet Cicely, 121
Swida, 119
Swine's Cress, 62
Sycamore, 87
Symphoricarpos, 169
Symphytum, 145
Syringa, 142

Tamus, 201
Tanacetum, 180
Tansy, 180
Taraxacum, 190
Tare, 94
Tassel Pondweed, 194
TAXACEAE, 51
Taxus, 51
Tea Tree, 148
Teasel, 172
Teucrium, 164
Thale Cress, 68
Thalictrum, 56
THELYPTERIDACEAE, 48
Thelypteris, 48
Thistle, 181, 182
Thlaspi, 63
Thorn-apple, 149
Thrift, 139
Thyme, 160
THYMELAEACEAE, 114
Thymus, 160
Tilia, 81
TILIACEAE, 81

Timothy Grass, 223
Toadflax, 150
Toothwort, 156
Torilis, 121
Tormentil, 102
Touch-me-not, 86
Tragopogon, 186
Traveller's Joy, 52
Treacle Mustard, 67
Tree Mallow, 82
Trefoil, 91
Trichophorum, 207
Trifolium, 91
Triglochin, 192
Trigonella, 91
Tripleurospermum, 179
Trisetum, 221
Trollius, 51
Turnip, 59
Turritis, 66
Tussilago, 174
Tutsan, 70
Twayblade, 202
Typha 206
TYPHACEAE, 206

Ulex, 89
ULMACEAE, 132
Ulmus, 132
UMBELLIFERAE, 119
Umbilicus, 111
Urtica, 132
URTICACEAE, 131
Utricularia, 157

Vaccinium, 138
Valerian, 171
Valeriana, 171
VALERIANACEAE, 170
Valerianella, 170
Verbascum, 149
Verbena, 158
VERBENACEAE, 158
Vernal Grass, 224
Veronica, 153
Vervain, 158
Vetch, 95
Vetchling, 96
Viburnum, 169
Vicia, 94
Vinca, 142
Viola, 69
VIOLACEAE, 69
Violet, 69

Viscum, 119
Vulpia, 215

Wahlenbergia, 165
Wallflower, 67
Wall Pennywort, 111
Wall Rue, 47
Walnut, 133
Wart Cress, 62
Water Blinks, 77
Watercress, 66
Water Dropwort, 124
Waterlily, 56
Water Milfoil, 117
Water Parsnip, 124
Water Pepper, 129
Water Plantain, 191
Water Purslane, 114
Water Rocket, 66
Water Starwort, 118
Wayfaring Tree, 169
Weasel-snout, 150
Weld, 68
Whin, 89
Whitebeam, 109
Whitlow-grass, 64
Whorl Grass, 217
Wild Angelica, 126
Wild Basil, 161
Wild Liquorice, 94
Willow, 135
Willow-herb, 115
Winter-Cress, 65
Wintergreen, 138
Winter Heliotrope, 175
Wood Pimpernel, 140
Woodruff, 167
Woodrush, 198
Wood Sage, 164
Wood Sanicle, 120
Wood Sorrel, 86
Wormwood, 180
Woundwort, 162

Yarrow, 179
Yellow Archangel, 162
Yellow-Cress, 66
Yellow-rattle, 155
Yellow-wort, 143
Yew, 51
Yorkshire Fog, 222

Zannichellia, 194
ZANNICHELLIACEAE, 194

236